Medical Intelligence Unit

Maternal Liver Disease

Jamal A. Ibdah, MD, PhD
Department of Internal Medicine
Division of Gastroenterology and Hepatology
University of Missouri
Columbia, Missouri, USA

CRC Press
Taylor & Francis Group
Boca Raton London New York

CRC Press is an imprint of the
Taylor & Francis Group, an **informa** business

Maternal Liver Disease

Medical Intelligence Unit

First published 2012 by Landes Bioscience

Published 2021 by CRC Press
Taylor & Francis Group
6000 Broken Sound Parkway NW, Suite 300
Boca Raton, FL 33487-2742

ISBN 13: 978-1-58706-657-3 (hbk)

Visit the Taylor & Francis Web site at
http://www.taylorandfrancis.com

and the CRC Press Web site at
http://www.crcpress.com

Library of Congress Cataloging-in-Publication Data

Maternal liver disease / [edited by] Jamal A. Ibdah.
p. ; cm. -- (Medical intelligence unit)
Includes bibliographical references and index.
ISBN 978-1-58706-657-3 (alk. paper)
I. Ibdah, Jamal A., 1956- II. Series: Medical intelligence unit (Unnumbered : 2003)
[DNLM: 1. Liver Diseases. 2. Pregnancy Complications. WQ 240]

618.3'6--dc23

2012027811

Dedication

This book is dedicated with love to my family

To my sons Malik, Khalid, and Zain
Who have been the light of my life

To my wife Robin
Who has been the love of my life

To my mother
Who has been the inspiration of my life

About the Editor...

DR. JAMAL A. IBDAH obtained his medical degree from the University of Jordan followed by a PhD in Biochemistry from the Medical College of Pennsylvania. He underwent residency training in Internal Medicine at Wake Forest University followed by a fellowship in Gastroenterology and Hepatology at Washington University. He received board certification in both Internal Medicine and Gastroenterology from the American Board of Internal Medicine. Currently, he is a Professor of Internal Medicine and Medical Pharmacology and Physiology at the University of Missouri-Columbia. He serves as Director of the Division of Gastroenterology and Hepatology, Raymond E. and Vaona H. Peck Chair in Research, Senior Associate Dean in the School of Medicine, and Director of the Institute for Clinical and Translational Science. Dr. Ibdah serves on the editorial board of *World Journal of Gastroenterology*, *World Journal of Gastrointestinal Oncology*, and *Journal of Cancer Therapy*, and has served as the senior editor for the *Journal of Cardio Metabolic Syndrome*. In addition, Dr. Ibdah has served as a member on the NIH Clinical Research and Review Committee, hepatobiliary pathophysiology National Institutes of Health (NIH) study section, and on several NIH special emphasis panels in clinical research.

Dr. Ibdah has devoted his research career to team science and his work exemplifies the translational approach from the laboratory to bedside. For instance, Dr. Ibdah brought to the bench a clinical observation of an association between liver disease in pregnant women and pediatric fatty acid oxidation disorders, which led to a discovery that contributed to the understanding of the pathogenesis of a disease that has been long considered to be mysterious (*New England Journal of Medicine*, 1999). Subsequently, his work has focused on elucidating the cellular mechanisms involved in the pathogenesis of these disorders. Work in his laboratory has uncovered the molecular basis for a link between fatty acid oxidation disorders and sudden unexpected infant death (*Journal of Clinical Investigation*, 2001). His work has resulted in important contributions that have crossed boundaries among disciplines. All studies were conducted in collaborative, multidisciplinary environments with a team approach that involved investigators in Pediatrics, Obstetrics, Endocrinology, Biochemistry, Comparative Medicine, and other disciplines. His work has been published in top-tier journals in the fields of Medicine, Pediatrics, Obstetrics, and research journals. Dr. Ibdah work has been funded by NIH, March of Dimes, and American Liver Foundation.

CONTENTS

EDITOR

Jamal A. Ibdah
Department of Internal Medicine
Division of Gastroenterology and Hepatology
University of Missouri
Columbia, Missouri, USA
Chapters 2, 4, 6-8

CONTRIBUTORS

Jerry Angdisen
Department of Internal Medicine
Division of Gastroenterology
and Hepatology
University of Missouri
Columbia, Missouri, USA
Chapter 7

Murtaza Arif
Department of Internal Medicine
Division of Gastroenterology
and Hepatology
University of Missouri
Columbia, Missouri, USA
Chapter 3

Yannick Bacq
Service d'Hépatogastroentérologie
Hôpital Trousseau
Tours, France
Chapter 1

Dilip Bearelly
Department of Internal Medicine
Division of Gastroenterology
and Hepatology
University of Missouri
Columbia, Missouri, USA
Chapter 6

Abhishek Choudhary
Department of Internal Medicine
Division of Gastroenterology
and Hepatology
University of Missouri
Columbia, Missouri, USA
Chapters 7, 8

Paul A. Dawson
Division of Gastroenterology
Department of Internal Medicine
Wake Forest University
School of Medicine
Winston-Salem, North Carolina, USA
Chapter 5

Hazem T. Hammad
Department of Internal Medicine
Division of Gastroenterology
and Hepatology
University of Missouri
Columbia, Missouri, USA
Chapter 2

Ghassan M. Hammoud
Department of Internal Medicine
Division of Gastroenterology
and Hepatology
University of Missouri
Columbia, Missouri, USA
Chapters 6-8

Kenneth L. Koch
Section on Gastroenterology
Department of Internal Medicine
Wake Forest University
School of Medicine
Winston-Salem, North Carolina, USA
Chapter 4

Gretchen Koontz
Department of Obstetrics
and Gynecology
Wake Forest University
School of Medicine
Winston-Salem, North Carolina, USA
Chapter 6

David C. Merrill
Aurora Medical Group
and Fetal Medicine
Milwaukee, Wisconsin, USA
Chapter 6

Girish Mishra
Division of Gastroenterology
Department of Internal Medicine
Wake Forest University
School of Medicine
Winston-Salem, North Carolina, USA
Chapter 5

Dennis Ng
Department of Internal Medicine
Division of Gastroenterology
and Hepatology
University of Missouri
Columbia, Missouri, USA
Chapter 3

William M. Outlaw
Section on Gastroenterology
Department of Internal Medicine
Wake Forest University
School of Medicine
Winston-Salem, North Carolina, USA
Chapter 4

R. Sidney G. Smith
Department of Internal Medicine
Division of Gastroenterology
and Hepatology
University of Missouri
Columbia, Missouri, USA
Chapter 2

PREFACE

Pregnancy is a very exciting time, but can be challenging to both the pregnant patient and the caring physician when complicated by a disorder, considering the impact on the health of both the mother and the fetus. This book provides a comprehensive and up-to- date review of liver disorders during pregnancy that will benefit both health care providers and researchers. *Maternal Liver Disease* represents a single source that contains detailed knowledge of clinical manifestations of maternal liver disorders, approach to diagnosis and management, clinical outcomes, as well as a detailed description of the latest scientific discoveries and research findings on pathogenesis of liver disease in pregnancy. Some of these disorders, until recently, have been considered mysterious and many of the recommendations, based on recent discoveries, can be lifesaving to the pregnant woman and her fetus. *Maternal Liver Disease* provides a novel contribution to the field through its comprehensive clinical and scientific focused approach.

Liver diseases that occur during the course of pregnancy are either secondary to conditions not specific to pregnancy or disease states which are specific to pregnancy. This book addresses these disorders in eight chapters divided into four sections. The first section, which contains Chapter 1, provides a review of physiological changes and alterations in liver function tests that occur during normal pregnancy. Comprehensive understanding of these physiological changes is necessary for the management of liver diseases.

The second section, which contains Chapters 2 and 3, provides a detailed review of maternal liver disorders not specific to pregnancy. These include viral hepatitis, autoimmune hepatitis, gall stone and biliary disease, hepatic vein thrombosis, primary biliary cirrhosis, hepatic adenoma, Wilson disease and portal hypertension. While most liver diseases not specific to pregnancy do not have a deleterious effect on the pregnancy itself, special precautions are advised and recommendations are provided based on recent literature.

The third section, which includes Chapters 4-7, is dedicated to maternal liver disease unique to pregnancy. In Chapter 4, the authors discuss hyperemesis gravidarum (HG), the most severe form of illness within the spectrum of nausea and vomiting of pregnancy. HG affects millions of pregnant women annually, imparts significant cost to society, and places mother and her unborn child at risk. Intrahepatic cholestasis of pregnancy (ICP) is discussed in Chapter 5. ICP is a reversible form of cholestasis that usually develops in the third trimester of pregnancy.

In Chapter 6, the authors discuss hypertensive disorders of pregnancy, which remain a devastating disease for both the mother and the fetus. These disorders include preeclampsia and HELLP (hemolysis, elevated liver enzymes, and low platelets) syndrome, which remain a major cause of maternal and neonatal mortality and morbidity. Acute fatty liver of pregnancy (AFLP) is discussed in Chapter 7. AFLP is a serious maternal illness occurring in the third trimester of pregnancy with significant perinatal and maternal mortality. Until recently, it has been considered a mysterious illness. Recent advances in understanding the pathogenesis of AFLP and their impact on the offspring born to women who develop AFLP are discussed.

The fourth section, which contains Chapter 8, is dedicated to liver transplantation in pregnancy. Liver transplantation has improved the lives of patients with decompensated acute and chronic liver disease. The authors provide detailed information on the challenges of pregnancy in liver transplant recipients and recent recommendations for close monitoring and follow up.

Jamal A. Ibdah, MD, PhD
Department of Internal Medicine
Division of Gastroenterology and Hepatology
University of Missouri
Columbia, Missouri, USA

Section I

The Normal Liver

Chapter 1

The Liver in Normal Pregnancy

Yannick Bacq*

Abstract

Changes in value of certain serum liver function tests occur during normal pregnancy and an understanding of these physiological changes is necessary for the management of liver diseases. Because of the phenomenon of hemodilution, the albumin level decreases as early as the first trimester. Serum alanine and aspartate aminotransferase activity levels remain within the normal limits established for nonpregnant women. Measurement of serum aminotransferase levels thus remains the most useful test for the routine diagnosis of liver diseases during pregnancy. Serum total and free bilirubin concentrations are lower in pregnant women than in nonpregnant controls during all three trimesters, as are concentrations of conjugated bilirubin during the second and third trimesters. Serum alkaline phosphatase activity increases in late pregnancy, due both to the production of the placental isoenzyme and to the increase in bone isoenzyme. Consequently, measurement of serum alkaline phosphatase levels is not a suitable test for the diagnosis of cholestasis during pregnancy. Serum gamma-glutamyl transferase activity levels decrease during the second and third trimesters, and serum 5'nucleotidase activity increases slightly in the second and third trimesters. Serum total bile acid concentrations during pregnancy are not different compared to levels in nonpregnant women. Measurement of serum bile acids may be useful for the diagnosis of cholestasis, especially when serum aminotransferase levels are within normal limits. In summary, according to the clinical setting, an increase in values of serum aminotransferase activity, serum GGT activity, serum bilirubin or fasting total bile acid concentrations during pregnancy should be considered pathologic.

Introduction

The pregnant woman experiences physiological changes to support fetal growth and development.[1] The levels of estrogens (estradiol) and progesterone increase progressively during pregnancy. These sex hormones have effects on hepatic metabolic, synthesis, and excretory functions.[2] The biliary excretion of bromosulfophthalein decreases during late pregnancy,[3,4] and the clearance of some compounds that are secreted into bile may therefore be impaired.[5] The phenomenon of hemodilution secondary to the increase in plasma volume decreases the serum protein concentrations. Consequently, certain changes in values of liver function tests occur during normal pregnancy. In this chapter, we review the liver-related clinical and biochemical changes that occur during normal pregnancy, with emphasis on liver function tests that are used for the management of liver diseases in pregnancy.

*Yannick Bacq—Service d'Hépatogastroentérologie, Hôpital Trousseau, 37044 Tours CEDEX, France. Email: bacq@med.univ-tours.fr

Liver-Related Symptoms and Physical Examination in Pregnancy

Nausea and vomiting are common symptoms of early pregnancy and occur in over half of all pregnant women.[6,7] The symptoms frequently persist throughout the day, although this condition was traditionally named "morning sickness".[7] By contrast, *Hyperemesis gravidarum*, usually defined by severe vomiting beginning in early pregnancy and requiring hospitalization, is much less frequent. As discussed in Chapter 4 by Outlaw et al, *Hyperemesis gravidarum* may be associated with liver involvement.[8] Nausea or vomiting occurring during the second or third trimester should be considered pathologic and should require investigations including the measurement of serum aminotransferase activity.[9] It should be noted that jaundice and generalized pruritus should never be considered normal features in pregnancy. Vascular spiders and palmar erythema are commonly associated with chronic liver disease and pregnancy. In one published study conducted in the US, vascular spiders were found in 14% of white women by the second month of pregnancy and in 66% by the ninth month of pregnancy.[10] The frequency of these vascular spiders was noted to be lower in black women with 8% occurring in the fourth month of pregnancy and 14% in the ninth month. In this study, about three-quarters of the women had lost these vascular spiders by the seventh week after delivery. In the same study, palmar erythema was observed in 63% of the white women and in 35% of the black women. By the time of the postpartum consultation, palmar erythema had faded in all but 9% of the women.[10] During pregnancy, these cutaneous vascular changes are not usually associated with hepatic dysfunction but may be related to sex steroids circulating in the blood over a long period. Physical examination of the liver is normal although it is difficult in late pregnancy because of the expanding uterus. In the postpartum period, after normal delivery the liver and spleen may be palpable because of ptosis of these organs.[11]

Hepatobiliary Ultrasonography in Pregnancy

Ultrasonography of the liver and biliary tract is widely used in the management of liver diseases outside pregnancy and is safe during pregnancy. In normal pregnancy, ultrasonographic examination reveals no dilatation of the biliary tract, but fasting gallbladder volume and residual volume after contraction are increased.[12-16] The lithogenic or cholesterol saturation index of bile increases during pregnancy, which is considered as a cholelithogenic state.[5,17] Indeed, biliary sludge frequently occurs during pregnancy but is generally asymptomatic and often disappears spontaneously after delivery.[18] Gallstones are much less frequent (2% in an Italian study, 12% in a Chilean study) and may be associated with biliary pain.[18,19] In the absence of suggestive symptoms, systematic ultrasound examination of the gallbladder as an extension of the routine pelvic ultrasound is not justified because silent stones in pregnant women need no treatment.[20]

Blood Volume and Hemodymamics in Pregnancy

Plasma volume increases steadily from the sixth to the 36th week of gestation by about 50%. The red cell volume also increases, but the increase is moderate (about 20%) and delayed. Consequently, the total blood volume increases with hemodilution. The hematocrit decreases by the 24th week and becomes stable. Plasma and red cell volume decrease rapidly after delivery.[21,22] This phenomenon of hemodilution should be kept in mind during interpretation of all serum concentrations during pregnancy.[23]

Cardiac output increases (about 40%) until the second trimester, then decreases and normalizes near term. Absolute hepatic blood flow remains unchanged, but the percentage of cardiac output to the liver decreases.[24-26]

Serum Protein and Lipid Levels in Normal Pregnancy

Serum albumin levels decrease during the first trimester, and this decrease becomes more accentuated as the pregnancy advances.[27] The decrease in serum concentration is explained by the hemodilution phenomenon. Indeed, the intravascular mass of albumin has been found to be normal in pregnancy and the rates of synthesis or catabolism are unaltered in normal pregnancy compared to controls.[28] By contrast, there is an increase in serum concentration of some proteins such as alpha2-macroglobulin, alpha1-antitrypsin, and ceruloplasmin.[29] Fibrinogen and most coagulation factors (II, VIII, IX and XII) increase when protein S levels decrease and fibrinolysis is inhibited.[30,31] These physiological changes in hemostasis limit bleeding during delivery but are associated with an increased risk of thromboembolism during pregnancy and the post-partum period.[30,31] Any change in the prothrombin time, which is widely used in the routine evaluation of chronic and acute liver failure, should be considered pathologic.

Serum cholesterol, triglyceride and phospholipid concentrations increase in late pregnancy.[32-35] This hyperlipidemia is a result of the metabolic adaptation to the pregnant state, saving glucose for the fetus.[36] Except when a pregnant woman is suffering from acute pancreatitis, measurement of serum lipid concentrations is rarely useful during pregnancy.[37] Serum concentrations of total globulins are unaffected by pregnancy.[34]

Serum Liver Function Tests in Normal Pregnancy

As outside pregnancy, serum liver function tests are essential in the management of liver diseases during pregnancy. Routine liver function tests usually include total and conjugated bilirubin, aminotransferases, alkaline phosphatase, and prothrombin time. In addition, gamma-glutamyltransferase (GGT) or 5'nucleotidase may be used to confirm the hepatobiliary origin of increased levels of alkaline phosphatase.[38] Measurement of serum bile acid concentrations may be useful for the management of cholestasis, especially during pregnancy.[39] However, this test is not easily available and this limits its utilization.[40] Awareness of the physiological changes in liver function tests is indispensable for the interpretation of these test values during pregnancy. Table 1 summarizes the changes in liver function tests during normal pregnancy compared to nonpregnant women.

Table 1. Changes in serum liver function tests during normal pregnancy

	Pregnancy Trimester		
Tests	**First**	**Second**	**Third**
Albumin	<	<	<
ALT	=	=	=
AST	=	=	=
Total bilirubin	<	<	<
Alkaline phosphatase	=	= or >	>
GGT	=	<	<
5'-nucleotidase	=	= or >	= or >
Total bile acids (fasting state)	=	=	=
Prothrombin time	=	=	=

= : no or slight changes, > : increase, < : decrease compared to non-pregnant women.

Bilirubin Concentrations

Total bilirubin concentrations are decreased during all three trimesters of pregnancy.[27,34,41] Free bilirubin concentrations have also been found to be lower in pregnant women than in nonpregnant controls during all three trimesters, as are concentrations of conjugated bilirubin during the second and third trimesters.[27] Hemodilution could at least be partly responsible for the decrease in bilirubin concentration because albumin is the protein that transports bilirubin.

Aminotransferase Activity

Measurement of serum alanine aminotransferase (ALT) and aspartate aminotransferase (AST) activity levels is the most useful test for the routine diagnosis of liver diseases. The effects of pregnancy in serum ALT and AST activity levels are somewhat controversial. In a few studies, a slight increase in ALT and/or AST activity has been found during the third trimester.[3,34,41-44] However, in the majority of published studies, serum ALT and AST activity levels do not change during pregnancy or remain within the normal limits established in nonpregnant women.[3,45-54] In one study, we found a slight increase in ALT during the second trimester of pregnancy compared to nonpregnant women, but all the values remained below the upper normal limit.[27] We did not have any explanation for this slight increase in ALT and only during the second trimester in pregnant women compared to nonpregnant controls.[27] An increase in ALT or AST levels during labor might be due to contractions of the uterine muscle.[55,56] Thus, it should be emphasized that serum AST or ALT activity values above the upper normal limit before labor should be considered pathologic and should lead to further investigations.[57]

Alkaline Phosphatase Activity

Serum alkaline phosphatase activity levels increase in late pregnancy, mainly during the third trimester.[27] By contrast, serum alkaline phosphatase levels have been found to be lower in oral contraceptive users.[34] This increase during pregnancy is not due to an increase in the hepatic isoenzyme but rather largely due to the production of the placental isoenzyme.[58-66] During the third trimester, there is also an increase in the production of the bone isoenzyme as documented by an increase in its serum level up to six weeks post delivery.[67-69] These findings document that the measurement of serum alkaline phosphatase activity is not a suitable test for the diagnosis of cholestasis during late pregnancy and post-partum.

Gamma-Glutamyl Transferase Activity

Serum GGT activity has usually been considered to be normal during pregnancy.[44,53,70-72] When we compared serum GGT activity levels between pregnant women and nonpregnant women not taking oral contraception we did not find difference in the first trimester.[27] However, we found a slight but significant decrease in GGT activity levels in the second and third trimesters in pregnant women (see Table 1). A significant decrease in serum GGT activity was also found in late pregnancy compared with early pregnancy in women with morning sickness.[41] On the other hand, women suffering from viral hepatitis in early pregnancy have higher levels of serum GGT activity compared to women with viral hepatitis in late pregnancy, and the same has been observed in women taking oral contraceptives.[73] These findings suggest that sex hormones may inhibit hepatic synthesis of GGT, leading to a decrease in serum GGT activity in late pregnancy.

5'Nucleotidase Activity

In two studies without control groups, no change was found in serum 5'nucleotidase activity.[74,75] In one study serum 5'nucleotidase activity was slightly higher in the second and third trimesters, compared with controls.[27] A slight increase during the third trimester was also observed in two other studies.[48,49] This slight physiological increase in serum 5'nucleotidase during late pregnancy is a limitation for its use in the diagnosis of cholestasis.

Bile Acid Concentrations

Increases in serum concentrations of certain bile acids have been reported during pregnancy, and it has been suggested that pregnancy could be associated with sub-clinical cholestasis.[41,76-79] These changes in serum bile acids were in fact minimal and observed mainly in the postprandial state. Moreover, some of these studies did not include a control group. By contrast, in a prospective study fasting total bile acid concentrations measured during each trimester of pregnancy did not significantly differ from the control group.[27] In clinical practice, when a woman experiences pruritus during pregnancy, the measurement of serum bile acid concentration may be useful for the diagnosis of cholestasis, especially when routine liver function tests are still within normal limits.[79,80]

Histological Studies

Standard and ultrastructural examination of the liver during normal pregnancy reveals no or minimal nonspecific abnormalities.[81-85]

Conclusion

During normal pregnancy, except for alkaline phosphatase and 5'nucleotidase, most values of serum routine liver function tests remain below the upper normal limits established in nonpregnant women. Thus, when liver disease is clinically suspected in a pregnant woman, any increase in serum ALT and AST activity levels, serum GGT activity, serum bilirubin or fasting total bile acid concentrations should be considered pathologic, and should prompt further evaluation.

References

1. Blackburn ST, Loper DL. Maternal, fetal, and neonatal physiology. A clinical perspective. Philadelphia: WB Saunders, 1992.
2. Van Thiel DH, Gavaler JS. Pregnancy-associated sex steroids and their effects on the liver. Semin Liver Dis 1987; 7(1):1-7.
3. Tindall VR, Beazley JM. An assessment of changes in liver function during normal pregnancy-using a modified bromsulphthalein test. J Obstet Gynaecol 1965; 72(5):717-737.
4. Ylöstalo P. Liver function in hepatosis of pregnancy and preeclampsia with special reference to modified bromsulphtalein tests. Acta Obstet Gynecol Scand 1970; 49(suppl 4):1-53.
5. Everson GT. Liver problems in pregnancy: Distinguishing normal from abnormal hepatic changes. Medscape Womens Health 1998; 3(2):3.
6. Klebanoff MA, Koslowe PA, Kaslow R et al. Epidemiology of vomiting in early pregnancy. Obstet Gynecol 1985; 66(5):612-616.
7. Vellacott ID, Cooke EJA, James CE. Nausea and vomiting in early pregnancy. Int J Gynecol Obstet 1988; 27(1):57-62.
8. Morali GA, Braverman DZ. Abnormal liver enzymes and ketonuria in hyperemesis gravidarum. A retrospective review of 80 patients. J Clin Gastroenterol 1990; 12(3):303-305.
9. Bacq Y, Riely CA. The liver in pregnancy. In: Schiff ER, Sorrel MF, Maddrey WC, eds. Schiff's diseases of the liver. 9th ed. Philadelphia: Lippincott Williams and Wilkins, 2003:2:1435-1457.
10. Bean WB, Cogswell R, Dexter M et al. Vascular changes of the skin in pregnancy: Vascular spiders and palmar erythema. Surg Gynecol Obstet 1949; 88:739-752.
11. Berris B. The incidence of palpable liver and spleen in the post-partum period. Can Med Assoc J 1966; 95(25):1318-1319.
12. Braverman DZ, Johnson ML, Kern F. Effects of pregnancy and contraceptive steroids on gallbaldder function. N Engl J Med 1980; 302(7):362-364.
13. Everson GT, McKinley C, Lawson M et al. Gallbladder function in the Human Female: Effect of the ovulatory cycle, pregnancy, and contraceptive steroids. Gastroenterology 1982; 82(4):711-719.
14. Stauffer RA, Adams A, Wygal J et al. Gallbladder disease in pregnancy. Am J Obstet Gynecol 1982; 144(6):661-664.

15. Ylöstalo P, Kirkinen P, Heikkinen J et al. Gall bladder volume and serum bile acids in cholestasis of pregnancy. Br J Obstet Gynecol 1982; 89(1):59-61.
16. Kirkinen P, Ylöstalo P, Heikkinen J et al. Gallbladder function and maternal bile acids in intrahepatic cholestasis of pregnancy. Eur J Obstet Reprod Biol 1984; 18(1-2):29-34.
17. Kern F, Everson GT, DeMark B et al. Biliary lipids, bile acids, and gallbladder function in the Human Female. Effects of pregnancy and the ovulatory cycle. J Clin Invest 1981; 68(5):1229-1242.
18. Maringhini A, Ciambra M, Baccelliere P et al. Biliary sludge and gallstones in pregnancy: Incidence, risk factors, and natural history. Ann Intern Med 1993; 119(2):116-120.
19. Valdivieso V, Covarrubias C, Siegel F et al. Pregnancy and cholelithiasis: Pathogenesis and natural course of gallstones diagnosed in early post-partum. Hepatology 1993; 17(1):1-4.
20. Scott LD. Gallstone disease and pancreatitis in pregnancy. Gastroenterol Clin North Am 1992; 21(4):803-815.
21. Lund CJ, Donovan JC. Blood volume during pregnancy. Am J Obstet Gynecol 1967; 98(3):393-403.
22. Peck TM, Arias F. Hematologic changes associated with pregnancy. Clin Obstet Gynecol 1979; 22(4):785-798.
23. Goodlin RC. Serum albumin reserve for bilirubin binding during pregnancy in healthy women. Obstet Gynecol 1989; 73(6):1054.
24. Munnell EW, Taylor HC. Liver blood flow in pregnancy. Hepatic vein catheterization. J Clin Invest 1947; 26:952-956.
25. Rovinsky JJ, Jaffin H. Cardiovascular hemodynamics in pregnancy. III. Cardiac rate, stroke volume, total peripheral resistance, and central blood volume in multiple pregnancy. Synthesis of results. Am J Obstet Gynecol 1966; 95(6):787-794.
26. Laakso L, Ruotsalainen P, Punnonen R et al. Hepatic blood flow during late pregnancy. Acta Obstet Gynecol Scand 1971; 50(2):175-178.
27. Bacq Y, Zarka O, Brechot J-F et al. Liver Function tests in normal pregnancy: A prospective study of 103 pregnant women and 103 matched controls. Hepatology 1996; 23(5):1030-1034.
28. Honger PE. Albumin metabolism in normal pregnancy. Scand J Clin Lab Invest 1968; 21(1):3-9.
29. Ganrot PO. Variation of the concentration of some plasma proteins in normal adults, in pregnant women and in newborns. Scand J Clin Lab Invest 1972; 29(suppl 124):83-88.
30. Bremme KA. Haemostatic changes in pregnancy. Best Pract Res Clin Haematol 2003; 16(2):153-168.
31. Hellgren M. Hemostasis during normal pregnancy and puerperium. Semin Thromb Hemost 2003; 29(2):125-130.
32. Svanborg A, Vikrot O. Plasma lipid fraction, including individual phospholipids at various stages of pregnancy. Acta Med Scand 1965; 178(5):615-630.
33. Knopp RH, Warth MR, Carrol CJ. Lipid metabolism in pregnancy. I. Changes in lipoprotein triglyceride and cholesterol in normal pregnancy and the effects of diabetes mellitus. J Reprod Med 1973; 10(3):95-101.
34. Knopp RH, Bergelin RO, Wahl PW et al. Clinical chemistry alterations in pregnancy and oral contraceptive use. Obstet Gynecol 1985; 66(5):682-690.
35. Kaabachi N, Fellah H, Abdelmoula J et al. Evolution des lipides au cours de la grossesse. J Gynecol Obstet Biol Reprod 1992; 21(5):544-548.
36. Dejager S, Turpin G. Hyperlipidémie au cours de la grossesse. Presse Med 1996; 25(37):1839-1845.
37. Chen CP, Wang KG, Su TH et al. Acute pancreatitis in pregnancy. Acta Obstet Gynecol Scand 1995; 74(8):607-610.
38. Sotil EU, Jensen DM. Serum enzymes associated with cholestasis. Clin Liver Dis 2004; 8(1):41-54.
39. Glantz A, Marschall HU, Mattsson LA. Intrahepatic cholestasis of pregnancy: Relationships between bile acid levels and fetal complication rates. Hepatology 2004; 40(2):467-474.
40. Egerman RS, Riely CA. Predicting fetal outcome in intrahepatic cholestasis of pregnancy: Is the bile acid level sufficient? Hepatology 2004; 40(2):287-288.
41. Järnfelt-Samsioe A, Eriksson B, Waldenström J et al. Serum bile acids, gamma-glutamyltransferase and routine liver tests in emetic and nonemetic pregnancies. Gynecol Obstet Invest 1986; 21(4):169-176.

42. Elliott JR, O'Kell RT. Normal clinical chemical values for pregnant women at term. Clin Chem 1971; 17(3):156-157.
43. Cerutti R, Ferrari S, Grella P et al. Behaviour of serum enzymes in pregnancy. Clin Exp Obstet Gynecol 1976; 3:22-24.
44. Salgo L, Pal A. Variation in some enzymes in amniotic fluid and maternal serum during pregnancy. Enzyme 1989; 41(2):101-107.
45. Carter J. Liver function in normal pregnancy. Aust NZ Obstet Gynaecol 1990; 30(4):296-302.
46. Moniz CF, Nicolaides KH, Bamforth FJ et al. Normal reference ranges for biochemical substances relating to renal, hepatic, and bone function in fetal and maternal plasma throughout pregnancy. J Clin Pathol 1985; 38(4):468-472.
47. Borglin NE. Serum transaminase activity in uncomplicated and complicated pregnancy and in newborns. J Clin Endocrinol 1958; 18(8):872-877.
48. Kater RMH, Mistilis SP. Obstetric cholestasis and pruritus of pregnancy. Med J Aust 1967; 25(13):638-640.
49. Kreek MJ. Cholestasis of pregnancy and during ethynil estradiol administration in the human and rat. In: Salhanick HA, Kipnis DM, Vande Wiele RL, eds. Metabolic effects of gonadal hormones and contraceptive steroids. New York-London: Plenum Press, 1969:40-58.
50. Rosing U, Kallner A. Serum cholic and chenodeoxycholic acids during pregnancy and puerperium in normal and preeclamptic women. Acta Obstet Gynecol Scand 1986; 65(1):11-13.
51. Theisen R, Jackson R, Morrissey J et al. Serum enzymes in normal and toxemic pregnancy. Obstet Gynecol 1961; 17(2):183-190.
52. Stone ML, Lending M, Slobody LB et al. Glutamic oxaloacetic transaminase and lactic dehydrogenase in pregnancy. Am J Obstet Gynecol 1960; 80(1):104-107.
53. Walker FB, Hoblit DL, Cunninggham FG et al. Gamma glutamyl transpeptidase in normal pregnancy. Obstet Gynecol 1974; 43(5):7450-749.
54. West M, Zimmerman HJ. Lactic dehydrogenase and glutamic oxaloacetic transaminase in normal pregnant women and newborn children. Am J Med Sci 1958; 235(4):443-447.
55. Meade BW, Rosalki SB. Serum enzyme activity in normal pregnancy and the newborn. J Obstet Gynaecol Br Commonw 1963; 70:693-700.
56. Reichard H, Wiqvist N, Yllner S. Chloroform anesthesia in obstetrics its effect of the serum activity of ornitine carbamyl transferase, glutamic oxaloacetic transaminase, and glutamic pyruvic transaminase. Acta Obst Gynaecol Scand 1960; 39:661-674.
57. Bernuau J. Complications hépatiques de la grossesse. Rev Prat 2003; 53(17):1889-1893.
58. McMaster Y, Tennant R, Clubb JS et al. The mechanism of the elevation of serum alkaline phosphatase in pregnancy. J Obstet Gynaecol Br Commonw 1964; 71:735-739.
59. Birkett DJ, Done J, Neale FC et al. Serum alkaline phosphatase in pregnancy: An immunological study. Br Med J 1966; 1:1210-1212.
60. Hunter RJ, Pinkerton JHM, Johnston H. Serum placental alkaline phosphatase in normal pregnancy and preeclampsia. Obstet Gynecol 1970; 36(4):536-546.
61. Elder MG. Serum heat-stable alkaline phosphatase levels in normal and abnormal pregnancy. Am J Obstet Gynecol 1972; 113(6):833-837.
62. Fishman WH, Bardawil WA, Habib HG et al. The placental isoenzyme of alkaline phosphatase in sera of normal pregnancy. Am J Clin Pathol 1972; 57(1):65-74.
63. Romslo I, Sagen N, Haram K. Serum alkaline phosphatase in pregnancy. I. A comparative study of total, L-phénylanine-sensitive and heat-stable alkaline phosphatase at 56°C and 65°C in normal pregnancy. Acta Obstet Gynecol Scand 1975; 54(5):437-442.
64. Sussman HH, Bowman M, Lewis JL. Placental alkaline phosphatase in maternal serum during normal and abnormal pregnancy. Nature 1968; 218(139):359-360.
65. Zuckerman H, Sadovsky E, Kallner B. Serum alkaline phosphatase in pregnancy and puerperium. Obstet Gynecol 1965; 25(6):819-824.
66. Merrett JD, Hunter RJ. Serum heat stable alkaline phosphatase levels in normal and abnormal pregnancies. J Obstet Gynaecol Br Commonw 1973; 80(11):957-965.
67. Adeniyi FA, Olatunbosun DA. Origins and significance of the increased plasma alkaline phophatase during normal pregnancy and preeclampsia. Br J Obstet Gynaecol 1984; 91(9):857-862.

68. Valenzuela GJ, Munson LA, Tarbaux NM et al. Time-dépendent changes in bone, placental, intestinal, and hepatic alkaline phosphatase activities in serum during human pregnancy. Clin Chem 1987; 33(10):1801-1806.
69. Rodin A, Duncan A, Quartero HWP et al. Serum concentrations of alkaline phosphatase isoenzymes and osteocalcin in normal pregnancy. J Clin Endocrinol Metab 1989; 68(6):1123-1127.
70. Crépin G, Querleu D, Delahousse G et al. Etude de la gamma-glutamy-transpeptidase et du rapport IgA sur transferrine chez la femme enceinte (et sous contraception orale) et le nouveau-né. J Gynecol Obstet Biol Reprod 1979; 8(7):625-630.
71. Lum G, Gambino SR. Serum gamma-glutamyl transpeptidase activity as an indicator of disease of the liver, pancreas, or bone. Clin Chem 1972; 18(4):358-362.
72. Moniz C, Nicolaides KH, Keys D et al. Gamma-glutamyl transferase activity in fetal serum, maternal serum, and amniotic fluid during gestation. J Clin Pathol 1984; 37(6):700-703.
73. Combes B, Shore GM, Cunningham FG et al. Serum gamma-glutamyl transpeptidase activity in viral hepatitis: Suppression in pregnancy and by birth control pills. Gastroenterology 1977; 72(2):271-274.
74. Alvi MH, Amer NA, Sumerin I. Serum 5-nucleotidase and serum sialic acid in pregnancy. Obstet Gynecol 1988; 72(2):171-174.
75. Seitanidis B, Moss DW. Serum alkaline phosphatase and 5'nucleotidase levels during normal pregnancy. Clin Chim Acta 1969; 25(1):183-184.
76. Carter J. Serum bile acids in normal pregnancy. Br J Obstet Gynaecol 1991; 98(6):540-543.
77. Fulton IC, Douglas JG, Hutchon DJR et al. Is normal pregnancy cholestatic? Clin Chim Acta 1983; 130(2):171-176.
78. Heikkinen J, Mäentausta O, Ylöstalo P et al. Changes in serum bile acid concentrations during normal pregnancy, in patients with intrahepatic cholestasis of pregnancy and in pregnant women with itching. Br J Obstet Gynaecol 1981; 88(3):240-245.
79. Lunzer M, Barnes P, Byth K et al. Serum bile acid concentrations during pregnancy and their relationship to obstetric cholestasis. Gastroenterology 1986; 91(4):825-829.
80. Bacq Y. Hépatopathies au cours de la grossesse. Gastroenterol Clin Biol 2001; 25(8-9):791-798.
81. Abide JK. Liver function in pregnancy utilising bromosulfophtalein. Obstet Gynecol 1966; 27(4):544-548.
82. Antia FP, Bharadwaj TP, Watsa MC et al. Liver in normal pregnancy, preeclampsia, and eclampsia. Lancet 1958; 2(7050):776-778.
83. Gonzalez-Angulo A, Aznar-Ramos R, Marquez-Monter H et al. The ultrastructure of liver cells in women under steroid therapy. I. Normal pregnancy and trophoblastic growth. Acta Endocrinol 1970; 65(2):193-206.
84. Ingerslev M, Teilum G. Biopsy studies on the liver in pregnancy. I. Normal histological features of the liver as seen on aspiration biopsy. Acta Obstet Gynecol 1945; 25(3):339-351.
85. Perez V, Gorodisch S, Casavilla F et al. Ultrastructure of human liver at the end of normal pregnancy. Am J Obstet Gynecol 1971; 110(3):428-431.

Section II

Liver Disease Not Specific to Pregnancy

CHAPTER 2

Spectrum of Maternal Liver Disease

Hazem T. Hammad, R. Sidney G. Smith and Jamal A. Ibdah*

Abstract

Liver dysfunction can be seen in up to 3% of pregnant women.[1] Liver diseases that occur during the course of pregnancy are either secondary to conditions not specific to pregnancy or certain disease states which are specific to pregnancy. Liver diseases unique to pregnancy include maternal liver disease associated with hyperemesis gravidarum in the first trimester, preeclamptic liver disease and HELLP syndrome in the second and third trimesters, acute fatty liver in pregnancy in the third trimester, and intrahepatic cholestasis of pregnancy in the third trimester. Some of the liver diseases specific to pregnancy can cause significant morbidity and mortality to the mother and the fetus. Liver diseases not specific to pregnancy that will be discussed include viral hepatitis, autoimmune hepatitis, gallstones and biliary disease, hepatic vein thrombosis, primary biliary cirrhosis, hepatic adenoma, Wilson disease, and portal hypertension. While most liver diseases not specific to pregnancy may not have a deleterious effect on the pregnancy itself, special precautions are advised.

Introduction

This chapter provides a brief overview of liver diseases unique to pregnancy, all of which are discussed in depth in subsequent chapters. With respect to illnesses not unique to pregnancy, this chapter provides a comprehensive overview, including the effect of these illnesses on the outcome of pregnancy, as well as the treatment options. Topics reviewed include pregnancy and viral hepatitis, autoimmune hepatitis, gallstones and biliary tract disorders, hepatic and portal vein thrombosis, hepatocellular adenoma, primary biliary cirrhosis, primary sclerosing cholangitis, Wilson disease, and portal hypertension. Pregnancy after liver transplantation is also discussed.

Liver Diseases Unique to Pregnancy

A brief description is provided below. These pregnancy-induced liver diseases will be discussed in great detail in subsequent chapters.

Hyperemesis Gravidarum

Hyperemesis gravidarum (HG) is the most severe form of illness within the spectrum of nausea and vomiting of pregnancy (NVP). HG affects millions of pregnant women annually, imparts significant cost to society, and places mother and her unborn child at risk. The nausea and vomiting characteristic of HG presents early in pregnancy and usually terminates by the 20th gestational week. Liver disease, usually consisting of mild serum transaminase elevation, occurs in almost 50% of patients with HG. While multiple risk factors have been proposed,

*Corresponding Author: Jamal A. Ibdah—Department of Internal Medicine, Division of Gastroenterology and Hepatology, University of Missouri, Columbia, Missouri, USA. Email: ibdahj@health.missouri.edu

Maternal Liver Disease, edited by Jamal A. Ibdah. ©2012 Taylor & Francis Group, LLC

the etiology of HG and the maternal liver disease seen in HG remains unclear. Current investigations underscore the potential roles of starvation injury, placental release of inflammatory cytokines, and impairment of fatty acid oxidation in the pathogenesis of the maternal liver disease observed in HG.

Intrahepatic Cholestasis of Pregnancy

Intrahepatic cholestasis of pregnancy (ICP), also known as obstetric cholestasis, is one of the most common causes of liver disease in pregnancy. ICP is a reversible form of cholestasis that usually develops in the third trimester of pregnancy and is associated with itching, elevated serum levels of bile acids, abnormal liver function tests, and mild changes in liver histology. These symptoms persist until delivery, after which the condition rapidly resolves. ICP is largely a clinical diagnosis, and is made by a suggestive history and exclusion of other causes of cholestasis. While ICP is considered a benign disease with a good prognosis for the mother, ICP is associated with poor fetal outcomes and perinatal mortality in untreated cases as high as 11 to 20%. The etiology of ICP is unclear, but undoubtedly involves genetic and hormonal factors. Recent genetic studies have implicated gene mutations in the liver canalicular phospholipid transporter, MDR3, however these mutations can account for only a small percentage of ICP cases. Current evidence suggests a possible abnormality in the metabolism of progesterone and bile acids leading to elevated maternal serum bile acids and impaired transport of bile acids across the placenta from the fetal to the maternal circulation. However, it is not known how this increased fetal bile acid burden leads to the complications of perinatal mortality, preterm delivery, abnormal intrapartum fetal heart rate rhythm, and meconium passage. Ursodeoxycholic acid (UDCA) is currently the most efficacious medical intervention. UDCA therapy relieves maternal symptoms, reduces serum liver enzymes and, more importantly, may also protect the fetus from bile acid toxicity and improve fetal survival. Due to the increased risk of stillbirth in ICP, most experienced centers advocate delivery by the 37th week of gestation.

Preeclamptic Liver Disease and HELLP Syndrome

Hypertensive disorders of pregnancy remain a devastating disease for both the mother and the fetus. Preeclampsia and HELLP syndrome, two disorders specific to pregnancy, remain a major cause of maternal and neonatal mortality and morbidity worldwide. Unfortunately, the diagnosis is not always straightforward as patients present with varying symptoms and degrees of severity. The laboratory evaluation remains the main diagnostic modality.

The liver is a major target with often devastating consequences. The pathogenesis of hepatic damage in cases of severe preeclampsia and HELLP syndrome is not well understood. The pathogenesis of liver damage is difficult to study as liver enzyme evaluations and maternal symptoms often correlate poorly with pathological findings. The characteristic liver pathology seen is that of dense fibrin accumulation leading to hemorrhage, necrosis and wide areas of liver infarction. The only treatment available remains to be expeditious delivery. Corticosteroids may be of potential benefit, although this remains unclear. Other specific treatment options such as liver transplantation and hepatic artery embolization are only done in the most extreme circumstances in cases of liver rupture.

Despite extensive research and advances in technology and antepartum care, there still remain many unanswered questions. The potential molecular mechanisms mitigating the liver involvement in this disease are discussed, but at present are only speculative. Areas of research continue to focus on the etiology of the disease and markers of disease progression. Future research will likely focus on the genetic risk and etiology of the disease. These conditions will be discussed in detail in Chapter 6 by Bearelly et al.

Acute Fatty Liver of Pregnancy

Acute fatty liver of pregnancy (AFLP) is a serious maternal illness occurring in the third trimester of pregnancy with significant perinatal and maternal mortality. Till recently, it has been considered a mysterious illness. This condition will be discussed in great detail in Chapter 7 by Choudhary et al. Recent advances in understanding the pathogenesis of AFLP documented a causative association between carrying a fetus with a defect in fatty acid oxidation and development of AFLP. The impact of these recent advances on the offspring born to women who develop AFLP can be life-saving. Screening for a genetic defect in the newborn can lead to a diagnosis of life threatening illness and allow early treatment before symptoms. It also would allow genetic counseling in subsequent pregnancies. The molecular basis and underlying mechanism for this unique fetal-maternal interaction causing maternal liver disease will be discussed in Chapter 7.

Liver Diseases Not Specific to Pregnancy

While evaluating liver dysfunction during pregnancy, it is important to emphasize that most liver abnormalities in pregnancy occur secondary to conditions not specific to pregnancy and such conditions should be considered along with those unique to pregnancy. These diseases can be pre-existing or ensue during the course of pregnancy. Pregnancy is generally uncommon in patients with severe chronic liver disease due to the associated altered sex hormone metabolism, resulting in anovulation.[2] Consequences of chronic liver disease on pregnancy include risk of worsening liver disease (increase in portal pressure, variceal bleeding and portosystemic encephalopathy) as well as risks to the developing fetus (premature labor, stillbirth, and low birthweight). Below is a discussion of liver disorders not specific to pregnancy with emphasis on special considerations for these disorders during the course of pregnancy, delivery and postpartum period.

Viral Hepatitis

Viral hepatitis is discussed in great detail in Chapter 3 by Ng and Arif. An overview is provided below.

Hepatitis A

Hepatitis A virus (HAV), a hepatotropic RNA virus, is transmitted via the fecal oral route and often comes to the attention of the public in the setting of food borne outbreaks. In the United States, the overall incidence of HAV is 9.1 per 100,000, while the incidence of acute HAV during pregnancy is believed to be very rare (less than 1:1000). The average incubation period is 30 days (range, 15 to 49 days). Overall, the presentation is nonspecific and can resemble other viral illnesses, with prodromal symptoms, such as malaise, anorexia, nausea and vomiting. In the vast majority of adults, jaundice usually develops within a few weeks of these symptoms and is associated with marked elevations in liver transaminases. Diagnosis of HAV is made by detection of antibodies (IgM) to HAV. Treatment consists largely of supportive care, with special care to maintain adequate hydration and nutrition. Although rare, fulminant hepatic failure can occur with acute HAV and patients should be followed closely. HAV does not lead to chronic infection, although 10 to 15% of symptomatic patients can have prolonged or relapsing course lasting up to 6 months.

The clinical presentation, disease course, treatment and sequela are similar to those of non-pregnant patients. Acute HAV in the setting of pregnancy does not appear to be teratogenic. Generally, HAV infection is well tolerated by both mother and fetus, although a recent report showed high risk of maternal complications and preterm labor with HAV infection during the second and third trimesters.[3] In patients with chronic hepatitis C, acute infection with HAV can lead to fulminant hepatic failure.[4] Thus; it is recommended that all patients with chronic HCV, irrespective of pregnancy status, be vaccinated against HAV. The currently available inactivated vaccines are safe and effective in pregnancy. Additionally, HAV immunoglobulin can be used for passive immunization if given within 2-3 weeks of a known exposure. Both inactivated vaccine against HAV and the post exposure immunoglobulin prophylaxis are safe in pregnancy.[5] Intrauterine transmission is rare and it is known that HAV can also be transmitted via the fecal oral route at the time of birth or in the perinatal period. HAV infection in mothers is not a contraindication to breastfeeding.

Hepatitis B

Hepatitis B virus (HBV) is a double-stranded, enveloped DNA virus, and a member of the Hepadnaviridae family. Around 350 million people are infected with HBV worldwide, including 1.5 million in the United States. HBV is highly infectious with the main routes of transmission being parenteral, sexual and perinatal. Perinatal transmission represents the most common route in China and the Far East. The clinical presentation and course of the disease is similar to nonpregnant infected individuals. Based on the available data, it appears that acute HBV infection complicates one to two of every 1000 pregnancies.[6] Fortunately, it does not appear that acute infection has any more of a deleterious effect on the pregnant female or fetus than any other individual.[7] Likewise, being pregnant does not affect the course of acute infection or progression of chronic disease.

The diagnosis of acute HBV infection is established by the detection of hepatitis B surface antigen (HBsAg) in the serum. The acute infection is confirmed by the presence of IgM antibodies to hepatitis B core antigen (HBcAg). Hepatitis e (envelope) antigen is present in almost all acute infections and represents high infectivity. The clinical presentation of acute infection is similar to the nonpregnant patient, with nausea, vomiting, right upper quadrant pain and jaundice being common. Fever can also be present and accompany the typical marked elevation in transaminases.[8] Five to ten percent of infected individuals will progress to chronic infection. As long as HBsAg is present, the individual is infectious. As anti-HBe antibody develops, infectivity decreases.

The risk of vertical transmission of HBV to the fetus is an issue of great importance and is affected by the timing of acute infection during pregnancy, e antigen status of the mother, as well as the viral load. If the acute maternal infection occurs in the first trimester and resolves, the risk for neonatal infection is minimal. In contrast, an infection during the second and third trimesters poses a threat of 10 and 90%, respectively, for vertical transmission.[9] Although intrauterine infection has been reported, the intrapartum period is the time of greatest risk, as it appears that 95% of transmission occurs during that period.[10] Intrauterine infection is rare, risk factors include: maternal serum HBeAg positivity, history of threatened preterm labor and HBV in the placenta.[11] Without neonatal prophylaxis, a child born to a HBeAg positive mother has 70-90% chance of being infected, while a child born to a HBeAg negative mother has less than 10% risk for infection.[12] The younger the child is at the time of infection, the greater the risk for chronic infection. A child infected at birth, has a 90% chance to progress to chronic infection, while an infection between 7 and 12 months will progress to chronicity in 40% of cases and only 10-20% of cases if the infection occurred between 1 to 3 years of age.[13]

Due to concern over vertical transmission, prenatal screening for HBV is now standard in many countries in order to vaccinate unprotected patients, as use of the currently available vaccine is considered safe during pregnancy. Pregnant women who get exposed to HBV should receive HBV immunoglobulins (HBIG) within 72 hours and then get HBV vaccine series, with first dose administered within 7 days ideally. Should the pregnant patient be found to have chronic infection or become infected during pregnancy, the infant should receive both passive and active immunization, with HBIG and HBV vaccine, regardless of the patient's E antigen status.[9] With the above combination, the risk of transmission is greatly reduced (up to 95% reduction). In addition, there is an emerging role for use of antiviral therapy late in pregnancy to decrease the risk of perinatal transmission. A recent report by Xu et al showed significant reduction in HBV perinatal transmission when pregnant women with high HBV DNA received lamivudine during the third trimester. In this study, no safety concerns were noted in the lamivudine treated mothers or their infants.[14] Lamivudine, which is considered an FDA category C in pregnancy, has been shown to be associated with increased birth defects when used in the first trimester. Among the other available antiviral medications, tenofovir and telbivudine are pregnancy category B and could also be considered for use during pregnancy. The rest of the antiviral medications and interferon are considered category C and are not recommended. More data are needed to clarify the role of antiviral medications for HBV prophylaxis during pregnancy. Although HBV DNA has been detected in the colostrum of HBsAg positive mothers, breast-feeding, with proper immunoprophylaxis, poses no additional risk for the transmission of the hepatitis B virus to the infants.[15]

Hepatitis C

Hepatitis C virus (HCV) is an enveloped single-stranded RNA virus, and a member of the flaviviridae family. The main risk factor for transmission is injection drug use; other risk factors include blood transfusion before 1992, transfusion of clotting factors before 1987, occupational needle sticks, sexual and perinatal transmission. There are 170 million people with HCV infection worldwide, including 2 to 3 million in the United States. The prevalence of detectable HCV antibodies continues to be a major public health problem and chronic infection is currently the leading indication for liver transplantation in the United States.

HCV infection becomes chronic in a larger proportion of cases (up to 80%).[5] There has been significant decrease in the incidence of acute HCV infection since 1990. The overall incidence of acute infection during pregnancy is difficult to quantify, in part due to the fact that acute infection is largely asymptomatic. However, it appears that the incidence of chronic HCV infection in pregnant women is similar to that of the general population. Numerous studies have evaluated pregnant patients for antibody to HCV, finding ranges from 0.14 to 2.4%. Using averages from the largest of these studies shows 1.26% of pregnant patients positive for anti-HCV, with 60% having positive HCV-RNA.[16]

Similar to chronic HBV, pregnant females with chronic HCV, in the absence of cirrhosis, do not seem to be at increased risk for fetal complications and in general tolerate pregnancy well.[17] However, a recent population-based cohort study using Washington state birth records from 2003 to 2005 showed infants of HCV-positive mothers were more likely to be low birthweight, to need assisted ventilation, and to require neonatal intensive care unit admission.[18] Some studies have shown that there is a trend toward normalization of transaminases during the second and third trimesters of pregnancy, inspite of a concurrent rise in serum HCV-RNA above prepregnancy levels.[19] Following pregnancy, transaminases tend to return to near their prepregnancy values. Contrary to the above, there is evidence that pregnancy can have an adverse impact on both biochemical markers and histology. Thus, no definite conclusions can be made regarding the course of chronic HCV during pregnancy.

Unlike HBV infection in which vertical transmission is a major issue, HCV does not seem to be transmitted as easily. While transmission rates vary according to the study cited, in general rates in human immunodeficiency virus (HIV)-negative patients are less than 10% and probably closer to 4-7%.[16] The presence of viremia (with detectable HCV RNA) is necessary for vertical transmission. If the mother is anti-HCV positive but HCV RNA negative, there is usually no risk for transmission.[20] Maternal co-infection with HIV has been documented to be a major risk factor for vertical transmission; with rates being several fold higher than in HIV-negative patients (up to 36%).[21,22] Other risk factors for vertical transmission include higher maternal HCV viral titer, prolonged membrane rupture during labor (6 hours or longer) and invasive fetal monitoring during labor.[20]

In the absence of other indications, HCV infection alone does not represent sufficient grounds for a Cesarean section. Although HCV antibodies and HCV RNA have been detected in the colostrum of HCV infected mothers, breastfeeding appears to be safe.[23] Since passive transmission of IgG antibodies to HCV is present through the placenta, a detectable anti HCV antibodies in the newborn does not mean that a perinatal infection has occurred. Detectable anti-HCV antibodies in infants more than 18 months old or a detectable HCV RNA in infants 2-6 months old can define mother to infant transmission of the infection.[16,20] The outcome of the children infected perinataly is still not clear. Several studies have demonstrated spontaneous clearance of HCV by 6-24 months of age.[24,25]

To date, no vaccine has been successfully made for HCV and there is no role for prophylaxis in the setting of pregnancy. Routine prenatal screening is not recommended. However screening of women with risk factors can be considered.[16] The currently approved treatments, ribavirin and interferon alpha, are both contraindicated in pregnancy. Interferon alpha is contraindicated in children under the age of 2 years because of severe neurotoxicity and ribavirin is considered teratogenic. The recently introduced protease inhibitors (Bocepravir and Telaprevir) are only used in combination with ribavirin and interferon alpha, and therefore the combination is also contraindicated during pregnancy.[26]

Hepatitis D

Hepatitis D virus (HDV) is an RNA virus that is dependant upon the presence of HBV for replication. Individuals with chronic HBV who become infected with HDV appear to have a more rapidly progressive course and are at higher risk for developing cirrhosis. The diagnosis of HDV is established by serological testing. Vertical transmission is possible but preventable by using established practices to prevent HBV infection.[6]

Hepatitis E

Hepatitis E virus (HEV) is a non-enveloped, single-stranded RNA virus. Although rare in the United States, HEV is a common problem in many parts of the developing world or among travelers or immigrants from endemic areas. HEV, which is usually spread through contaminated drinking water, can occur in either sporadic or epidemic forms.[6] The incubation period is usually 15 to 60 days. HEV can be difficult to diagnose and can be confused on clinical and laboratory findings with acute fatty liver of pregnancy. The diagnosis of HEV requires the detection HEV in serum or stool by PCR or the detection of HEV IgM antibodies. The infection is usually self-limited and treatment is mainly supportive. Pregnant women appear to be especially susceptible to HEV infection.[27] Moreover, HEV infection during pregnancy, particularly in the last two trimesters, is associated with a high rate of fulminant hepatic failure (up to sixty-percent) with maternal death rates up to 31%.[28,29] The etiology of the severe clinical course of HEV during pregnancy is not well understood, but may be related to immunologic factors.[30] Although vertical transmission can occur with devastating consequences, to date there is no therapy to prevent transmission.[31,32]

Autoimmune Hepatitis

Autoimmune hepatitis (AIH) is a generally progressive, chronic hepatitis of unknown cause that occurs in children and young adults. It is characterized by loss of tolerance to hepatic autoantigens with resultant T-cell-mediated immune attack on liver antigens, leading to hepatic necroinflammation and fibrosis.[33] If left untreated, severe disease eventually leads to cirrhosis. The disease is characterized by elevated levels of aminotransferases, hypergammaglobulinemia and circulating serum autoantibodies. The prevalence in the North America and European Caucasian population is between 0.1-1.2 cases per 100,000.[34] As is the case with other autoimmune disorders, women are more commonly affected by autoimmune hepatitis than men and this frequently occurs during the childbearing years. Moreover, AIH has been reported to occur during pregnancy or in the postpartum period and should be considered in the patient who presents with new onset liver disease following pregnancy.[35] Prior studies have reported that there is an increased rate of infertility associated with autoimmune hepatitis.[36] Although the exact mechanism by which chronic liver disease and cirrhosis cause amenorrhea and anovulation, with resultant infertility, has not been completely elucidated, studies suggest hypothalamic-pituitary dysfunction.[37] Regardless of the mechanism, if AIH is treated successfully, normal menstrual cycles may return with the possibility of pregnancy.

The course of AIH during pregnancy is not completely clear, with data showing conflicting patterns of disease activity. However, in general it appears that the disease is milder during pregnancy, possibly as a result of the relative state of immuno-tolerance, which seems to be predominating during that time. In one study of 6 female patients (five with AIH and one with autoimmune sclerosing cholangitis) involving 14 pregnancies, liver tests returned to normal by month 4 in all but 2 pregnancies.[38] This normalization occurred despite the fact that azathioprine had been withdrawn. It is also notable that the steroid dose could be reduced in all patients who were using them. In the same study, 12 of the 14 pregnancies were complicated by flares in the postpartum period despite intensification of treatment. Moreover, a recent report described five new cases of severe AIH, which presented within 4 months of delivery.[39] Due to the frequent exacerbation of AIH after delivery, its occurrence must be anticipated, and conventional therapy must be resumed pre-emptively two weeks before anticipated delivery and maintained throughout the postpartum period.[33]

In regard to the outcome of pregnancy in patients with AIH, the fetal loss rate appears to be less than previously reported. In a review of 35 pregnancies, Heneghan et al reported fetal loss rate of 14%.[35] The cesarean section and prematurity rates in this study were also low at 3 and 6% respectively. In a more recent study of 42 pregnancies in 22 AIH patients, the risk of adverse pregnancy outcome was 26% and was highly associated with the presence of antibodies to SLA/LP and Ro/SSA.[40] A review of the reported cases in the literature from 1966 to 2004 showed that in 101 pregnancies in patients with AIH there were 47 flare-ups, 5 clinical improvements, and 45 stabilizations of the disease during pregnancy. There were two maternal deaths and 19 fetal deaths with a perinatal mortality of 4%.[41] Pregnancy appears to be safe in patients with well controlled AIH.[42] Patients need careful monitoring throughout the pregnancy, especially in the postpartum period because of relapses. Although azathioprine has been reported to cause birth defects in animal models, based on the data available to date in humans, it appears that the use of azathioprine during pregnancy is safe.[35]

Gallstones and Biliary Tract Disease

There are several changes that occur during the course of pregnancy that lead to the formation of biliary sludge and gallstones. Among these is a decline in the contractility of the gallbladder with resultant stasis and increase in residual volume.[43] It appears that this is likely due to hormonal changes, with progesterone having the greatest effect. In addition to changes in emptying, there is also an alteration in the content of bile, in the form of an estrogen-induced increase in cholesterol secretion, progesterone-induced reduction in bile acids secretion, and an increase in cholic acid component of bile along with decrease in the deoxycholic acid and chenodeoxycholic acid components. The overall effect of these changes appears to increase the cholesterol saturation of bile, making it more lithogenic.[44]

Although symptomatic biliary tract disease during pregnancy is rare, the incidence of gallstones ranges from 4.5-12% and increases with additional pregnancies.[43] Prepregnancy obesity and serum leptin were found to be strong risk factors for pregnancy-associated gallbladder disease.[45] During pregnancy, gallstones usually manifest as biliary pain, gallstone pancreatitis or less likely as acute cholecystitis. The management of symptomatic uncomplicated biliary tract disease is usually conservative, with intravenous hydration and antibiotics. Although this approach is largely successful, a more aggressive approach may be required in cases of severe disease such as acute cholecystitis, choledocholithiasis, cholangitis, or gallstone pancreatitis. Contrary to older studies, recent data show that cholecystectomy during pregnancy is not associated with a high rate of fetal demise.[46] Likewise, recent reports indicate that laparoscopic cholecystectomy is probably a safe option as well.[47,48] Irrespective of the surgical approach, surgery is considered to be safest during the second trimester of pregnancy.

Endoscopic management of biliary tract disease using endoscopic retrograde cholangiopancreatography (ERCP) is also a viable option. Two recent retrospective studies provide further evidence to support the safety of ERCP during pregnancy, provided that all possible measures are taken to minimize fetal radiation exposure and that the clinical situation warrants intervention.[49,50]

Hepatic Vein Thrombosis (Budd-Chiari Syndrome) and Portal Vein Thrombosis

It is well recognized that pregnancy induces a hypercoagulable state, which extends into the postpartum period, in part through an increase in certain clotting factors and a decrease in coagulation inhibitors.[51] In addition to the risk of lower extremity thrombosis with the chance of pulmonary embolism, both the hepatic and portal veins can also be affected.

Budd-Chiari syndrome (BCS) is a rare disease that is caused by the obstruction of hepatic venous outflow, leading to sinusoidal congestion, ischemic injury to the liver and portal hypertension. Its association with pregnancy has been well documented. Even though pregnancy itself can induce a hypercoagulable state, it is becoming increasingly evident that other predisposing conditions can co-exist, such as factor V Leiden mutation or prothrombin gene mutation. Thus, a thorough thrombophilia evaluation should be undertaken. The disease usually presents in the third trimester or the puerperium. In general, the outcomes of acute BCS in the setting of pregnancy are poor. In addition to anticoagulation treatment with low-molecular-weight heparin (LMWH), there is a recent case report of a surgical portacaval shunt being successfully placed in a patient during the second trimester of pregnancy, thus allowing for successful completion of the pregnancy.[52] Likewise, there is a case report of a successful liver transplantation for BCS in the setting of pregnancy.[53] However, it should be noted that although liver transplantation has been performed during pregnancy, the outcome for the fetus is poor.[54] After the diagnosis and treatment of BCS, pregnancy is not contraindicated in stable patients with well-controlled disease. However, fetal outcome remains relatively poor, which could be related to underlying prothrombotic disorders.[55]

Portal vein thrombosis (PVT) occurs rarely during pregnancy. It is usually caused by local or systemic factors. Local factors include malignant tumors, cirrhosis, and intra-abdominal infections. Systemic factors include hypercoagulable disorders such as myeloproliferative disorders, antiphospholipid syndrome and factor V Leiden. Treatment for acute PVT includes anticoagulation. A 2009 guideline by the American Association for the Study of Liver Diseases (AASLD) recommends giving anticoagulation therapy for at least 3 months. In pregnant patients with chronic PVT, screening for gastroesophageal varices and adequate prophylaxis for variceal bleeding should be provided.[56]

Pregnancy after Liver Transplantation

Patients with chronic and end-stage liver disease, as described previously, have a high incidence of amenorrhea, up to approximately 50%, in some series.[57] However, menstruation and sexual drive usually return after successful liver transplantation, making pregnancy possible. In the above referenced series, by 7 months after transplantation, 90% of the posttransplant premenopausal women had resumed menstruation. Furthermore, it appears that this can occur even in the presence of abnormal liver function tests.[58]

There are multiple case series in the literature of pregnancies in female liver transplant recipients, with 219 pregnancies having been reported to the National Transplantation Pregnancy Registry as of 2007.[59] These pregnancies should certainly be considered high-risk pregnancies and comprehensive medical care should be provided. With regard to the timing of pregnancy, it appears that pregnancies that occur closer to the time of transplantation are associated with a worse fetal outcome.[60] On the basis of these findings, it is generally recommended that pregnancy be delayed for at least 2 years after transplant to allow for the mother's medical status to stabilize.[61,62]

Even in patients delaying pregnancy for the recommended time post transplant, there continues to be a higher than normal incidence of intrauterine growth restriction, preterm delivery, and cesarean section rate. Maternal hypertension posttransplant has been well described and is in part linked to the immunosuppressive regimen, particularly cyclosporine.[63] Preeclampsia is also common, especially in patients with underlying renal dysfunction.[60] Some studies have suggested that this can be reduced with the use of a tacrolimus-based immunosuppressive regimen.[63] A review of the literature published in 2006 showed a 3% incidence of fetal malformations in children of transplanted mothers which is comparable to the incidence in the general population. Immunosuppressive drugs in utero seem to be safe and have not been linked to fetal malformations.[62] Due to concern over possible adverse outcomes related to exposure to medications, breastfeeding is not recommended.[63]

Hepatocellular Adenoma

Although overall hepatocellular adenomas are uncommon, there has clearly been an increase in their incidence with the introduction of oral contraceptives. The magnitude of this increase is not clear, with more recent studies showing a smaller increase than previously reported.[64] In addition to oral contraceptive use, there is also an association between pregnancy and adenomas which is likely secondary to increased levels of steroid hormones.[65] Although there is a small risk of malignant transformation, the more worrisome issue with adenomas is the risk of rupture, with potentially fatal consequences. A recent review found a mortality rate of 44 and 38% respectively for mother and fetus, when rupture occurred during pregnancy.[64] The risk of rupture is highest in the third trimester.[66]

Although there have been no large studies addressing the optimal management of adenomas in the setting of pregnancy, it is generally recommended that pregnancy be avoided in patients with large adenomas (>5 cm) or those who have experienced complications of adenomas in previous pregnancies. Pregnancy may be allowed in the case of a known small adenoma, after discussing the risk-benefit ratio with the patient, and provided that the pregnancy is closely monitored.[66] If an adenoma is discovered during pregnancy, management can be based, to some degree, on the initial size of the lesion. If less than 5 cm in size, the adenoma can be followed closely with serial ultrasounds. However, for a lesion that is enlarging or initially greater than 5 cm, surgery is recommended.[64] There are case reports of resection during pregnancy with good outcomes for both mother and fetus.[67] When resection is not feasible, radiofrequency ablation can be used.[66] Fujita et al reported a case of a pregnant patient with multiple adenomas that were treated by a combination of resection and radiofrequency ablation during her 18th gestational week with successful completion of pregnancy thereafter.[68] It should be noted that patients with adenomas should be followed closely in the postpartum period, as they also appear to be at risk for rupture during this time.[64]

Primary Biliary Cirrhosis and Primary Sclerosing Cholangitis

Overall, pregnancy in patients with primary biliary cirrhosis (PBC) is rare and there is little data available regarding the natural history of the disease if pregnancy does occur. Based on studies published in the late sixties, it appeared that the prognosis for both mother and fetus was poor.[69] More recent publications have in general shown that pregnancy does not cause deterioration in liver function and that fetal outcome is good.[70] However, it should also be noted that there are other reports describing worsening of liver function, especially when the diagnosis of PBC is made during pregnancy.[71] A recent report of nine pregnancies in six patients with ursodeoxycholic acid (UDCA)-treated PBC showed that all women remained asymptomatic during pregnancy with improvement in laboratory tests and no recurrence of pruritus.[72] In this series, a flare in disease with increase in liver biochemistries was observed 3 months postpartum. UDCA has been shown to be safe in pregnancy. The American Association for the Study of Liver Disease (AASLD) guidelines recommend screening for varices in the second trimester and treating portal hypertension with beta blockers if indicated.[73]

With regard to pregnancy in the setting of primary sclerosing cholangitis (PSC), a recent report compared 229 patients with PSC with 569 healthy controls and showed that fertility does not appear to be reduced in PSC patients and that successful pregnancies can be achieved by patients with PSC. There was no impairment of fetal outcome and no serious maternal complications were recorded.[74] Although there is no definite pattern of disease activity, jaundice seems to progress during pregnancy.[75] UDCA has been used safely and successfully to control pruritus during pregnancy in women with PSC.[74]

Wilson Disease

Wilson disease (WD) is a multisystem autosomal recessive disorder of hepatic copper transport with resultant inhibition of biliary copper excretion. This results in copper excess and deposition in the liver and brain. The disease occurs in 1:30,000 to 1:50,000 persons.[76] Studies on the effect of WD on pregnancy are limited to small case series. Untreated WD may adversely affect fertility due to hormonal fluctuations that can result in amenorrhea. A recent report showed higher rate of recurrent spontaneous abortions among women with WD who were untreated compared to women with WD who underwent treatment.[77]

Penicillamine, trientine, and zinc are drugs approved by the FDA for treatment of WD. Penicillamine acts by chelation and increases the excretion of copper in the urine. Trientine works similarly but is less effective than penicillamine. Zinc induces intestinal cells metallothionein that binds to copper and prevents transfer of copper into the blood. In a report of 26 pregnancies in 19 women with WD who were on zinc therapy throughout their pregnancy, zinc appeared to protect the health of the mother during pregnancy. With regard to the fetal outcome, two congenital defects occurred; one baby was born with a heart defect requiring surgery at 6 months, and a second baby was born with microcephaly.[78] In this study, zinc intake at a dose of 25 to 50 mg three times daily during pregnancy appeared to be safe with very minimal teratogenicity.

Penicillamine has been reported to cause teratogenicity in animals and humans.[79] Likewise, trientine has been to be teratogenic in animals.[80] It should be emphasized that treatment with chelation therapy should be maintained during pregnancy, as interruption of therapy has been associated with hemolytic episodes, hepatic insufficiency and maternal death. Both the mother and baby should be monitored while on chelation therapy.

Cirrhosis and Portal Hypertension

It is estimated that prevalence of cirrhosis in reproductive age women is 0.45 cases per 1,000 patients.[81] Although as described earlier, patients with severe liver disease frequently have menstrual irregularities and amenorrhea; pregnancy can occur but is thought to be rare. It should be noted that while in the western world portal hypertension is most commonly due to cirrhosis; noncirrhotic portal hypertension is more prevalent in the developing countries. In contrast to patients with cirrhosis, fertility has been shown to be near normal in these cases.[82] Pregnancy in patients with cirrhosis and portal hypertension requires close multidisciplinary medical supervision because of the complexities of these patients. Irrespective of whether underlying liver disease is present, pregnancy causes a number of physiologic changes in the body. Among these is a 40-50% increase in the circulatory volume and systemic vasodilatation.[83] These changes probably lead to worsening of portal hypertension with an increased risk of variceal bleeding, which appears to be greatest during the second trimester.[84]

Variceal bleeding complicates 18 to 38% of pregnancies in patients with cirrhosis and in up to 50% in those with known portal hypertension. In pregnant women with known varices, bleeding occurs in up to 78%.[85] In addition to the risk of variceal bleeding, pregnancy can lead to other complications such as worsening liver failure, portosystemic encephalopathy, pulmonary hypertension, rupture of a splenic artery aneurysm, or postpartum hemorrhage.[85,86] Overall, 30 to 50% of pregnant patients with portal hypertension have some complication and the mortality rate is 10 to 18%.[10] With respect to the fetus, there is also an increased risk of complications including spontaneous abortion and fetal wastage.[10] In a recent report, complications during pregnancy were associated with higher MELD scores at the time of conception.[87]

In light of this risk, it is recommended that all cirrhotic patients who are pregnant or considering pregnancy undergo upper endoscopy to evaluate for varices.[10] On the basis of the endoscopic findings, primary prophylaxis with a nonselective beta-blocker such as propranolol or nadolol (both are FDA category C) may be indicated, although there are risks of fetal bradycardia, hypoglycemia, and intrauterine growth retardation with their use.[83] If there is a history of prior variceal bleeding and poor liver function, it has been recommended that pregnancy be avoided.[10] If gastrointestinal bleeding does occur during pregnancy, in addition to medical therapy, upper endoscopy should be performed. It appears, based on limited data, that endoscopic band ligation is safe during pregnancy.[88,89] Vaginal delivery is usually safe, however, early assistance with forceps delivery or vacuum extraction should be considered to prevent excessive straining which could increase portal pressure.

Conclusion

The spectrum of liver disease that can arise during the course of pregnancy is broad and includes both disease states specific to pregnancy and those which simply occur during pregnancy. Regardless, liver disease can present in a dramatic fashion and incite significant apprehension in both the patient and the clinician. A thorough understanding of the physiologic changes and the hepato-biliary disease states that occur during pregnancy is required in order to appropriately diagnose and treat these patients. The information provided in this chapter should serve as a foundation upon which more knowledge can be added, in order to provide the best possible care for our pregnant patients with liver disease.

References

1. Ch'ng CL, Morgan M, Hainsworth I et al. Prospective study of liver dysfunction in pregnancy in Southwest Wales. Gut 2002; 51(6):876.
2. Varma RR. Course and prognosis of pregnancy in women with liver disease. Semin Liver Dis 1987; 7(1):59.
3. Elinav E, Ben-Dov IZ, Shapira Y et al. Acute hepatitis A infection in pregnancy is associated with high rates of gestational complications and preterm labor. Gastroenterology 2006; 130(4):1129-1134.
4. Vento S, Garofano T, Renzini C et al. Fulminant hepatitis associated with Hepatitis A virus superinfection in patients with chronic hepatitis C. N Engl J Med 1998; 338:286-290.
5. Magriples U. Hepatitis in pregnancy. Semin Perinatol 1998; 22(2):112-117.
6. Dinsmor MJ. Hepatitis in the obstetric patient. Infect Dis Clin North Am 1997; 11:77-91.
7. Wong S, Chan LY, Yu V et al. Hepatitis B carrier and perinatal outcome in singleton pregnancy. Am J Perinatol 1999; 16(9):485-488.
8. Knox TA, Olans LB. Liver disease in pregnancy. N Engl J Med 1996; 335(11):569-576.
9. Hunt MD, Sharara AI. Liver disease in pregnancy. Am Fam Physician 1999; 59:829-836.
10. Sandhu BS, Sanyal AJ. Pregnancy and liver disease. Gastroenterol Clin N Am 2003; 32:407-436.
11. Xu DZ, Yan YP, Choi BC et al. Risk factors and mechanism of transplacental transmission of hepatitis B virus: A case-control study. J Med Virol 2002; 67(1):20-26.
12. Beasley RP, Trepo C, Stevens CE et al. The e antigen and vertical transmission of hepatitis B surface antigen. Am J Epidemiol 1977; 105(2):94-98.
13. Beasley RP, Hwang LY, Lin CC et al. Incidence of hepatitis B virus infections in preschool children in Taiwan. J Infect Dis 1982; 146(2):198-204.
14. Xu WM, Cui YT, Wang L et al. Lamivudine in late pregnancy to prevent perinatal transmission of hepatitis B virus infection: a multicentre, randomized, double-blind, placebo-controlled study. J Viral Hepat 2009; 16(2):94-103.
15. Hill JB, Sheffield JS, Kim MJ et al. Risk of hepatitis B transmission in breast-fed infants of chronic hepatitis B carriers. Obstet Gynecol 2002; 99(6):1049-1052.
16. Roberts EA, Yeung L. Maternal-Infant transmission of hepatitis C virus infection. Hepatology 2002; 36(5 Suppl 1):S106-S113.
17. Jabeen T, Cannon B, Hogan J et al. Pregnancy and pregnancy outcome in hepatitis C type 1b. QJM 2000; 93:597-601.
18. Pergam SA, Wang CC, Gardella CM et al. Pregnancy complications associated with hepatitis C: data from a 2003-2005 Washington state birth cohort. Am J Obstet Gynecol 2008; 199(1):38.e1.
19. Gervais A, Bacq Y, Bernuau J et al. Decrease in serum ALT and increase in serum HCV RNA during pregnancy in women with chronic hepatitis C. J Hepatol 2000; 32:293-299.
20. Mast EE, Hwang LY, Seto DS et al. Risk factors for perinatal transmission of hepatitis C virus (HCV) and the natural history of HCV infection acquired in infancy. J Infect Dis 2005; 192(11):1880-1889.
21. Zanetti AR, Tanzi E, Paccagnini S et al. Mother-to-infant transmission of hepatitis C virus. Lombardy Study Group on Vertical HCV Transmission. Lancet 1995; 345(8945):289.
22. Ferrero, Simone; Pietro Lungaro; Bianca Marisa Bruzzone et al. Prospective study of mother-to-infant transmission of hepatitis C virus: a 10-year survey (1990-2000). Acta Obstet Gynecol Scand 2003; 82(3):229-234.
23. Polywka S, Schroter M, Feucht HH et al. Low risk of vertical transmission of hepatitis C by breast milk. Clin Infect Dis 1999; 29:1327-1329.
24. Ceci O, Margiotta M, Marello F et al. High rate of spontaneous viral clearance in a cohort of vertical infected hepatitis C virus infants: What lies behind. J Hepatol 2001; 35:687-788.
25. Ketzinel-Gilad M, Colodner SL, Hadary R et al. Transient transmission of hepatitis C virus from mothers to newborns. Eur J Clin Microbiol Infect Dis 2000; 19:267-274.
26. Ghany MG, Nelson DR, Strader DB et al. An update on treatment of genotype 1 chronic hepatitis C virus infection: 2011 practice guideline by the American Association for the Study of Liver Diseases. Hepatology 2011; 54(4):1433-1444.
27. Khuroo MS, Teli MR, Skidmore S et al. Incidence and severity of viral hepatitis in pregnancy. Am J Med 1981; 70:252-255.

28. Khuroo MS, Kamili S. Etiology, clinical course and outcome of sporadic acute viral hepatitis in pregnancy. J Viral Hepat 2003; 10:61-69.
29. Boccia D, Guthmann JP, Klovstad H et al. High mortality associated with an outbreak of hepatitis E among displaced persons in Darfur, Sudan. Clin Infect Dis 2006; 42(12):1679-1684.
30. Pal, Rekha; Rakesh Aggarwal; Subhash R Naik et al. Immunological alterations in pregnant women with acute hepatitis E. J Gastroenterol Hepatol 2005; 20(7):1094-1101.
31. Khuroo MS, Kamili S, Jameel S. Vertical transmission of hepatitis E virus. Lancet 1995; 345:1025-1026.
32. Kumar A, Beniwal M, Kar P et al. Hepatitis E in pregnancy. Int J Gynaecol Obstet 2004; 85(3):240-244.
33. Manns MP, Czaja AJ, Gorham JD et al. Diagnosis and management of autoimmune hepatitis. Hepatology 2010; 51(6):2193-2213.
34. Manns MP, Strassburg CP. Autoimmune hepatitis: Clinical challenges. Gastroenterology 2001; 120:1502-1517.
35. Heneghan MA, Norris SM, O'Grady JG et al. Management and outcome of pregnancy in autoimmune hepatitis. Gut 2001; 48:97-102.
36. Steven MM, Buckley JD, Mackay IR. Pregnancy in chronic active hepatitis. QJM 1979; 192:519-531.
37. Cundy TF, Butler J, Pope RM. Amenorrhea in women with nonalcoholic chronic liver disease. Gut 1991; 32:202-206.
38. Buchel E, Van Steenbergen W, Nevens F et al. Improvement of autoimmune hepatitis during pregnancy followed by flareup after delivery. Am J Gastroenterol 2002; 97:3160-3165.
39. Samuel D, Riordan S, Strasser S et al. Severe autoimmune hepatitis first presenting in the early post partum period. Clin Gastroenterol Hepatol 2004; 2:622-624.
40. Schramm C, Herkel J, Beuers U et al. Pregnancy in autoimmune hepatitis: outcome and risk factors. Am J Gastroenterol 2006; 101(3):556-560.
41. Candia L, Marquez J, Espinoza LR. Autoimmune hepatitis and pregnancy: a rheumatologist's dilemma. Semin Arthritis Rheum 2005; 35(1):49-56.
42. Terrabuio DR, Abrantes-Lemos CP, Carrilho FJ et al. Follow-up of pregnant women with autoimmune hepatitis: the disease behavior along with maternal and fetal outcomes. J Clin Gastroenterol 2009; 43(4):350-356.
43. Everson GT. Gastrointestinal motility in pregnancy. Gastroenterol Clin North Am 1992; 21:751-776.
44. Scott LD. Gallstone disease and pancreatitis in pregnancy. Gastroenterol Clin North Am 1992; 21:803-815.
45. Ko CW, Beresford SA, Schulte SJ et al. Incidence, natural history, and risk factors for biliary sludge and stones during pregnancy. Hepatology 2005; 41(2):359-365.
46. Cosenza CA, Bahman S, Jabbour N et al. Surgical management of biliary gallstone disease during pregnancy. Am J Surg 1999; 178:545-548.
47. Al-Fozan H, Tulandi T. Safety and risks of laparoscopy in pregnancy. Curr Opin Obstet Gynecol 2002; 14:375-379.
48. Jelin EB, Smink DS, Vernon AH et al. Management of biliary tract disease during pregnancy: a decision analysis. Surg Endosc 2008; 22(1):54-60.
49. Kahaleh M, Hartwell GD, Arseneau KO et al. Safety and efficacy of ERCP in pregnancy. Gastrointest Endosc 2004; 60:287-292.
50. Tham TC, Vandervoort MD, Wong RC et al. Safety of ERCP during pregnancy. Am J Gastroenterol 2003; 98:308-311.
51. Goodrich MA, James EM, Baldus WP et al. Portal vein thrombosis associated with pregnancy. J Reprod Med 1993; 37:969-972.
52. Grant WJ, McCashland T, Botha JF et al. Acute Budd-Chiari syndrome during pregnancy: Surgical treatment and orthotopic liver transplantation with successful completion of the pregnancy. Liver Transpl 2003; 9:976-979.
53. Fickert P, Ramschak H, Kenner L et al. Acute Budd-Chiari syndrome with fulminant hepatic failure in a pregnant woman with factor V Leiden mutation. Gastroenterology 1996; 111:1670-1673.

54. Eguchi S, Yanaga K, Fujita F et al. Living-related right lobe liver transplantation for a patient with fulminant hepatic failure during the second trimester of pregnancy. Transplantation 2002; 73:1970-1971.
55. Rautou PE, Angermayr B, Garcia-Pagan JC et al. Pregnancy in women with known and treated Budd-Chiari syndrome: maternal and fetal outcomes. J Hepatol 2009; 51(1):47-54.
56. DeLeve LD, Valla DC, Garcia-Tsao G. Vascular disorders of the liver. Hepatology 2009; 49(5): 1729-1764.
57. Cundy TF, O'Grady JG, Williams R. Recovery of menstruation and pregnancy after liver transplantation. Gut 1990; 31:337-338.
58. Mass K, Quint EH, Punch MR. Gynecological and reproductive function after liver transplantation. Transplantation 1996; 62:476-479.
59. Armenti VT, Constantinescu S, Moritz MJ et al. Pregnancy after transplantation. Transplant Rev (Orlando) 2008; 22(4):223-240.
60. Jain AB, Reyes J, Marcos A et al. Pregnancy after liver transplantation with tacrolimus immuno-suppression: A single center's experience update at 13 years. Transplantation 2003; 76:827-832.
61. Nagy S, Bush MC, Berkowitz R et al. Pregnancy outcome in liver transplant recipients. Obstet Gynecol 2003; 102:121-128.
62. Dei Malatesta MF, Rossi M, Rocca B et al. Pregnancy after liver transplantation: report of 8 new cases and review of the literature. Transpl Immunol 2006; 15(4):297-302.
63. Armenti VT, Herrine SK, Radomski JS et al. Pregnancy after liver transplantation. Liver Transpl 2000; 6:671-685.
64. Cobey FC, Salem RR. A review of liver masses in pregnancy and a proposed algorithm for their diagnosis and management. Am J of Surgery 2004; 187:181-191.
65. Terkivatan T, de Wilt JH, de Man RA et al. Management of hepatocellular adenoma during pregnancy. Liver 2000; 20:186-187.
66. Noels JE, van Aalten SM, van der Windt DJ et al. Management of hepatocellular adenoma during pregnancy. J Hepatol 2011; 54(3):553-558.
67. Hill MA, Albert T, Zieske A et al. Successful resection of multifocal hepatic adenoma during pregnancy. South Med J 1997; 90:357-361.
68. Fujita S, Kushihata F, Herrmann GE et al. Combined hepatic resection and radiofrequency ablation for multiple hepatic adenomas. J Gastroenterol Hepatol 2006; 21(8):1351-1354.
69. Goh SK, Gull SE, Alexander GJ. Pregnancy in primary biliary cirrhosis complicated by portal hypertension. BJOG 2001; 108:760-762.
70. Olsson R, Loof L, Wallerstedt S. Pregnancy in patients with primary biliary cirrhosis. Liver 1993; 13:316-318.
71. Rabinovitz M, Appasamy R, Finkelstein S. Primary biliary cirrhosis diagnosed during pregnancy-does it have a different outcome. Dig Dis Sci 1995; 40:571-574.
72. Poupon R, Chretien Y, Chazouilleres O et al. Pregnancy in women with ursodeoxycholic acid-treated primary biliary cirrhosis. J Hepatol 2005; 42:418-419.
73. Lindor KD, Gershwin ME, Poupon R et al. Primary biliary cirrhosis. Hepatology 2009; 50(1):291-308.
74. Wellge BE, Sterneck M, Teufel A et al. Pregnancy in primary sclerosing cholangitis. Gut 2011; 60(8):1117.
75. Gossard A, Lindor K. Pregnancy in a patient with primary sclerosing cholangitis. J Clin Gastroenterol 2002; 35:353-355.
76. Lee NM, Brady CW. Liver disease in pregnancy. World J Gastroenterol 2009 28; 15(8):897-906.
77. Sinha S, Taly AB, Prashanth LK et al. Successful pregnancies and abortions in symptomatic and asymptomatic Wilson's disease. J Neurol Sci 2004 15; 217(1):37-40.
78. Brewer GJ, Johnson VD, Dick RD et al. Treatment of Wilson's disease with zinc. XVII: treatment during pregnancy. Hepatology 2000; 31(2):364-370.
79. Solomon L, Abrams G, Dinner M et al. Neonatal abnormalities associated with D-penicillamine treatment during pregnancy. N Engl J Med 1977; 296:54-55.
80. Tanaka H, Yamanouchi M, Imai S et al. Low copper and brain abnormalities in fetus from triethylene tetramine dihydrochloridetreated pregnant mouse. J Nutr Sci Vitaminol 1992; 38:545-554.

81. Russell MA, Craigo SD. Cirrhosis and portal hypertension in pregnancy. Semin Perinatol 1998; 22:156-165.
82. Kochhar R, Kumar S, Goel RC et al. Pregnancy and its outcome in patients with noncirrhotic portal hypertension. Dig Dis Sci 1999; 44:1356-1361.
83. Misra S, Sanyal AJ. Pregnancy in a patient with portal hypertension. Clin Liver Dis 1999; 3:147-162.
84. López-Méndez E, Avila-Escobedo L. Pregnancy and portal hypertension a pathology view of physiologic changes. Ann Hepatol 2006; 5(3):219-223.
85. Tan J, Surti B, Saab S. Pregnancy and cirrhosis. Liver Transpl 2008; 14(8):1081-1091.
86. Sigel CS, Harper TC, Thorne LB. Postpartum sudden death from pulmonary hypertension in the setting of portal hypertension. Obstet Gynecol 2007; 110(2 Pt 2):501-503.
87. Westbrook RH, Yeoman AD, O'Grady JG et al. Model for end-stage liver disease score predicts outcome in cirrhotic patients during pregnancy. Clin Gastroenterol Hepatol 2011; 9(8):694-699.
88. Starkel P, Horsmans Y, Geubel A. Endoscopic band ligation: A safe technique to control bleeding esophageal varices in pregnancy. Gastrointest Endosc 1998; 48:212-214.
89. Ghidirim G, Mishin I, Dolghii A et al. Prophylactic endoscopic band ligation of esophageal varices during pregnancy. J Gastrointestin Liver Dis 2008; 17(2):236-237.

CHAPTER 3

Viral Hepatitis in Pregnancy

Dennis Ng and Murtaza Arif*

Abstract

Worldwide, it is estimated that over 350 million individuals are carriers of chronic hepatitis B. It is generally accepted that at least 50% of them were acquired perinatally or in early childhood, especially in countries where hepatitis B is endemic. The majority of newborn infants who become infected go on to become chronic carriers of hepatitis B. The CDC estimates that 25% of adults who become chronically infected during childhood later on die from cirrhosis or liver cancer, making this a major worldwide public health problem in the world. Therefore, all efforts should be made to adhere to WHO/CDC guidelines (prenatal screening, immunoprophylaxis and possibly anti-viral therapy) to decrease the rate of transmission to the newborn. Vaccination has been shown to decrease the incidence and prevalence of hepatitis B, in addition to decreasing the incidence of liver cancer.

Pregnant women infected with hepatitis C are also frequently encountered in clinical practice. High risk women should be screened for hepatitis C. Although treatment for hepatitis C is contraindicated during pregnancy, it is important to refer patients back to see a liver specialist after delivery to discuss further options for therapy, because successful eradication of hepatitis C has been shown to stop progression of liver disease and improve survival.

Acute hepatitis A and E are not frequently encountered in developed countries. Because they are transmitted mainly via the fecal oral route, it is mainly seen in developing countries where sanitation is poor. Treatment remains supportive. Hepatitis E continues to enjoy a bad reputation because it has long been observed to have a high incidence and more severe course in pregnant women in some geographical areas where HEV is endemic.

Herpes simplex hepatitis is rare but notoriously dangerous because it is commonly a fatal condition. It has a predilection for the immunosuppressed and also pregnant women in their third trimester. Because of its rarity and atypical presentation, most patients go undiagnosed and die of acute liver failure. Early treatment with acyclovir can be lifesaving.

We include a table (Table 3) that summarizes the various viral hepatitis regarding the effect on pregnancy, perinatal transmission rate, prevention of perinatal transmission and breastfeeding.

Introduction to Viral Hepatitis in Pregnancy

This chapter reviews two of the most common liver diseases (hepatitis C and hepatitis B in pregnant women) being referred to a typical gastroenterology outpatient clinic. It is directed mostly towards health care providers involved in the care of pregnant patients and their newborn. As such, no attempts have been made to go into comprehensive detail about the pathophysiology of the disease. Interested readers may refer to several of the excellent textbooks on liver diseases, AASLD guidelines and review articles. Brief mention will be made regarding viral hepatitis not

*Corresponding Author: Murtaza Arif—Department of Internal Medicine, Division of Gastroenterology and Hepatology, University of Missouri, One Hospital Drive, Columbia, Missouri, USA. Email: arifm@health.missouri.edu

Maternal Liver Disease, edited by Jamal A. Ibdah. ©2012 Taylor & Francis Group, LLC

commonly seen in the US including hepatitis A, hepatitis E and herpes simplex hepatitis because of the potential severity of the disease when acquired during pregnancy. We especially emphasize the unique opportunity health care providers involved in the care of pregnant women have in screening patients for hepatitis B and C. Patients found to have hepatitis B and C can then be referred to a liver specialist for further education about healthy lifestyle to prevent progression of liver disease (smoking, drug and alcohol cessation) and possible options for treatment either during pregnancy (in some cases of hepatitis B) or after delivery (in some cases of hepatitis B and hepatitis C).

General Presentation and Management of Hepatitis

Liver test abnormalities and jaundice are rare in pregnant women and are seen in about 0.3-3 % of pregnancies. But when they do occur, they may portend serious liver diseases either due to pregnancy- associated acute liver diseases (e.g., preeclampsia associated liver disease, acute fatty liver disease of pregnancy and hemolysis elevated liver enzymes and low platelet syndrome (HELLP)) or an acute onset viral hepatitis (e.g., hepatitis A and hepatitis B). Other less common causes of elevated LFTs like Budd-Chiari syndrome and drug-induced liver disease can also occur. Patients with rapidly rising LFTs, vomiting, dehydration, elevated INR and change in mental status should be admitted to hospital urgently where an expedited work up can be done. The definitive treatment for pregnancy-specific acute liver diseases is mostly delivery of the baby; however, the management of viral hepatitis is mainly conservative using an appropriate drug for the particular hepatitis if indicated. For example, early institution of acyclovir in herpes simplex hepatitis and nucleoside/nucleotide analogues for acute hepatitis B with liver failure can be life-saving, thus emphasizing the need for early diagnosis.

Hepatitis B is commonly discovered because of the adoption of hepatitis B screening during the prenatal visit in the United States. Equally, patients at high risk for acquiring hepatitis C should be screened for this disease. When found positive for hepatitis B or C, the patient should be referred to a liver specialist to determine the status of hepatitis B or C, severity of liver disease, eligibility for treatment and to help to manage the disease postdelivery. It is important to screen high-risk patients for hepatitis C because for some young women, pregnancy may be the only time they come in contact with health care providers. Once diagnosed, we have the golden opportunity to educate them about the mode of transmission, ways to prevent progression of disease (drug and alcohol cessation) and liver cancer screening if required.

Hepatitis B

Epidemiology

Hepatitis B virus (HBV) is a DNA virus. Its importance lies in the fact that it is the most common hepatitis virus worldwide and is the leading cause of chronic liver disease, cirrhosis, liver cancer and death.

Worldwide, it is estimated that more than 350 million individuals are chronic carriers of hepatitis B.[1] The prevalence of hepatitis B varies from region to region. Hepatitis B prevalence can be divided into low prevalence (<2%) (North America, Western Europe), intermediate prevalence (2-7%) and high prevalence (>8%) areas (Sub-Saharan Africa, Asia excluding Japan, the Pacific basin, parts of Middle East and the Amazon basin).[1]

Still, in the United States, it is important to recognize the recent increase in immigrants from countries where hepatitis B is highly prevalent (e.g., China, regions of South East Asia) over the last few decades. Between 1994 and 2003, the US admitted about 40,000 individuals with chronic hepatitis B infection annually.[2] The local prevalence of hepatitis B in areas with a large immigrant population tends to mirror that of their country of origin. As an example, a study in 2009 found that the seroprevalence of HepBsAg among Asian Chinese in an urban Chinatown Internal Medicine practice in Chicago is about 11.1%.[3]

The 2009 AASLD Chronic Hepatitis B guidelines recommends screening high risk population to identify HBV-infected persons, including immigrants and adopted children from areas with high or intermediate prevalence rates.[4] In a recent abstract in AASLD 2010, Loo et al studied the adherence rate of screening for hepatitis B in patients with an Asian ethnic background in a large academic primary care practice from 1994-2009. The screening rate was at an abysmal 31%. This study underscores the importance of screening pregnant women at risk for hepatitis B and referral for further care if appropriate. Fortunately in the US, we adopt the universal screening policy for hepatitis B during the first trimester of pregnancy.

Importance of Hepatitis B on a Global Scale

Hepatitis B has 2 modes of transmission, vertically from an infected mother to the newborn during delivery (vertical transmission) and horizontally from an infected sexual partner, or household contact (horizontal transmission). Of the 350 million individuals chronically infected, it is generally accepted that at least 50% acquired their infection perinatally or in early childhood, especially in countries where HBV is endemic.[1] This is attributed to the high rates of HepBeAg positive infections in women of child bearing age in such parts of the world and the efficient transmission of infection to their newborns.

The risk of developing chronic hepatitis B is inversely related to the age of acquiring hepatitis B. The risk of developing chronic hepatitis B is as high as 90% in newborns of HepBeAg positive mothers, about 30% when acquired in infants and children < 5 years old, and less than 5% when acquired as an adult.[4] According to CDC statistics, about 25% of adults who become chronically infected during childhood later die from liver cancer or cirrhosis. Worldwide, chronic hepatitis B remains a major health threat; each year approximately 600,000 individuals die of complications such as acute liver failure, cirrhosis and hepatocellular carcinoma.[6] Therefore, there is a recognized need to prevent perinatal transmission of hepatitis B to decrease the global burden of disease.

Diagnosing Hepatitis B in Mother

Over the years, we have developed better serological and molecular tests to detect hepatitis B. The lower limit of quantification of HBV Virus HBV DNA via PCR method is now at <10 IU/mL.

Typically, HepBsAg, HepBsAb and HepBcTotalAb should be ordered initially. If HepBsAg is positive, one may proceed with HepBeAg, HepBeAb and HBV DNA PCR. A common mistake is to order the acute hepatitis panel which usually consists of HepBsAg, HepBcIgM Ab, HCV Ab and HepA IgM. While it is useful in diagnosing acute hepatitis, it is not helpful to determine the phase of the hepatitis B infection the patient is in. (i.e., immunotolerant phase, immunoactive phase, low replicative phase, reactivation phase, remission phase).

It is important to educate patients that hepatitis B is a lifetime disease, which will require regular blood work and surveillance for liver cancer if they are in the high risk category. At present, unlike hepatitis C, we are not able to cure hepatitis B because of the presence of cccDNA in the nucleus. Patients who are not eligible for treatment currently (e.g., immunotolerant phase) may become a treatment candidate later in life (when their hepatitis B transition into another phase e.g., immunoactive phase). Therefore it is important for perinatal health care providers to refer patients to liver specialists for follow-up after pregnancy. Treatment of hepatitis B has been shown to decrease morbidity and mortality of hepatitis B patient by decreasing the rate of progression to cirrhosis and its complications, decreasing the risk of developing liver cancer and preventing death in patients with liver failure.

Effect of Hepatitis B on Pregnancy

Data on the impact of chronic hepatitis B and acute hepatitis B infection on pregnancy outcome are limited and controversial. Most of the data are from South East Asian countries making generalization to other populations difficult. Wong et al, in a large study in 1999, found no differences in gestational age at delivery, birth weight, incidence of prematurity, neonatal jaundice, congenital anomalies or perinatal mortality when comparing HepBsAg positive women with non-infected controls.[7] Later, in 2010, an Israeli study analyzing 186, 619 deliveries in a population-based study found that HBV infection was an independent risk factor for poor pregnancy outcome.[8] In 2011, Elefsiniotis, in a prospective study, analyzed data from 102 women with HepBeAg negative chronic HBV infection and came to the conclusion that the presence of HBV-DNA in cord blood is significantly associated with spontaneous preterm birth in chronic HBV infected pregnant woman.[9]

Susceptible women, especially in areas where hepatitis B is endemic can develop acute hepatitis B. Acute hepatitis B must be distinguished from other types of acute viral hepatitis versus pregnancy-associated liver diseases. It appears that acute hepatitis does not increase mortality of the mother and has no teratogenic side effects. However a higher incidence of low birth weight infants and prematurity has been reported. HBV infection in early pregnancy has been associated with a 10% perinatal transmission rate but this rate increased substantially with HBV infection during the last trimester.[12]

Effect of Pregnancy on Hepatitis B

Little is known about the natural progression of the disease during pregnancy. Most women with HBV infection who do get pregnant will exhibit a variable but stable disease pattern during the course of pregnancy. For the few who do experience a re-activation or a flare of hepatitis B, the consequences may be serious, including progressing to acute liver failure and fetal demise. Because there is no good predictor of which patients will flare up, pregnant patients should have their LFTs checked every 3 months and more frequently if indicated.[11] As reported by ter Borg et al, chronic hepatitis B has also been known to flare up after delivery of baby, possibly due to the reconstitution of the maternal immune system.[13] Therefore, HBV infected women should continue to be monitored postpartum. Patients who exhibit severe re-activation or flare of hepatitis B, showing signs and symptoms of acute liver failure, should be started immediately on potent antiviral therapy to prevent further progression of disease.

Prevention of Perinatal Transmission of Hepatitis B

Vaccination of Mothers

The CDC recommends that all women be tested for hepatitis B during the initial prenatal visit, even if the patients have been vaccinated in the past or tested before in prior pregnancies.

If negative for hepatitis B, pregnant women at risk for hepatitis B (e.g., having more than one sexual partner during the previous 6 months, been evaluated or treated for an STD, recent or current injection drug use or having had a HepBsAg positive sex partner) should be vaccinated. Vaccines are usually given during time 0, 1 and 6 months. Pregnant women with compliance issues who cannot show up for the complete course of hepatitis B vaccines can be considered for an accelerated course of vaccine given at 0,1 and 4 months. In 2011, Sheffield et al showed that an accelerated course of hepatitis B vaccination schedule at 0, 1 and 4 months in high risk pregnancy women is effective, practical and well tolerated, with seroconversion (hepatitis B surface antibody of >10 mIU/mL) comparable to the traditional schedule, thus providing another strategy to decrease hepatitis B virus infection and potential transmission to the baby.[23]

Table 1. CDC recommendation for infant vaccination for hepatitis B

At birth

- If mother is HepBsAg negative, administer single antigen hepatitis B vaccine to all newborns >2 kg at birth before hospital discharge
- If mother is HepBsAg positive, administer single antigen hepatitis B vaccine and 0.5mL IM of hepatitis B immunoglobulin (HBIG) within 12 hours of birth
- If mother is HepBsAg unknown, administer single antigen hepatitis B vaccine within 12 hours of birth. Determined mothers HepBsAg ASAP and if HepBsAg positive, administer HBIG no later than age 1 week

After the birth dose

- The hepatitis series should be completed within 6 months of life per protocol.
- Infants born to HepBsAg positive mothers should be tested for HepBsAg and HepBsAb after completion of hepatitis B series at age 9-18 months

Vaccinations of Infants

CDC recommendations are listed in Table 1.[14]

Unlike hepatitis C, hepatitis B has a notoriously high rate of vertical transmission. Up to 90% of infants born to HepBeAg positive mothers become infected with hepatitis B. Infection is generally asymptomatic and the majority of these infants go on to become chronic carriers. HepBeAg positivity reflects the high replicative rate of this virus and is a marker of high viral load in the blood. These patients are highly infectious. High maternal viral load appears to increase the risk of transmission to the baby. It is widely accepted that the most transmissions occur during the perinatal period, hence the universal recommendation of hepatitis B vaccines and hepatitis B immunoglobulin (HBIG). Immunoprophylaxis with hepatitis B vaccines and HBIG are effective in preventing transmission of hepatitis B in more than 95% of cases. The remaining 5% of the cases are usually those infants who did not get the full regimen of vaccination or who were born to mothers with very high viral loads, particularly HBV DNA > 8log IU/mL.[15] These findings have led to the suggestion of instituting anti-viral therapy to mothers with very high viral load in the hope of reducing the level of viremia at the time of delivery (the time at which actual transmission is thought to occur).

Efficacy of Vaccination over Time

Vaccination against hepatitis B is highly successful and about 90-99% of children develop HepBsAb. However, HepBsAb levels decline over time and most children lose detectable antibody during the first 10-15 years after being immunized.[21] For example, in a 20 year follow-up study of long term protection against hepatitis B after newborn vaccination, Roznovsky et al found that persistence of protective HepBsAb was detected in 70%, 40% and 25% of children at 5, 10 and 15 years of age, respectively.[24] Immunocompetent persons who achieve anti-HBs concentrations > 10 mIU/mL after pre-exposure vaccination have virtually complete protection against both acute disease and chronic infection even if anti-HBs concentrations subsequently decline to <10 mIU/mL.[22] As such, the CDC recommends that immunocompetent persons found to have anti-HBs levels > 10mIU/mL after primary vaccinations series do not need further periodic testing to assess anti-HBs levels.[14]

Success Story of Vaccination

Hepatitis B virus is the first human virus proven to cause cancer. Epidemiological studies support that hepatitis B is indeed a cause of liver cancer and that vertical transmission is the driving force behind the carrier state. With about 700,000 deaths annually, liver cancer ranks as the 3rd most common cause of cancer death worldwide. The success of vaccination in preventing maternal to infant transmission of hepatitis B and the fact that hepatitis B causes the majority of the liver cancers worldwide made it increasingly apparent that a global hepatitis B vaccine program is needed. In 1992, R. Palmer Beasley successfully convinced the World Health Assembly to include HBV vaccines as the 7th immunogen in the WHO global childhood immunization program. Today HBV vaccines exist in 171 countries.[25]

Taiwan became the first country to adopt a nationwide hepatitis B universal vaccination program for infants in 1984. In a 2009 paper, Ni et al describes the Taiwan's 20 year experience after adopting the vaccination program. Vaccination coverage was about 97%. The seroprevalence of HepBsAg declined from 9.8% (prevaccination period) to 0.6% in children in Taipei City after 20 years of mass vaccination. The incidence of HCC in 6 to 9 year-olds declined from 0.52/100,000 for those born between 1974 and 1984 to 0.13/100,000 for those born between 1984 to 1986.[26] The same kind of trend was found in Thailand. In a 2010 paper, Tajiri et al reviewed the annual reports from a nationwide survey of childhood sold tumors during the 28 year period from 1981 to 2008. They showed a decrease in the number of patients with HBV-related HCC over the study period.[27]

Perinatal Hepatitis B Prevention Program (PHBPP)

All mothers who are tested positive for hepatitis B should be referred to the state's PHBPP and to a liver specialist for counseling and medical treatment if indicated. PHBPP is a state-run, CDC-funded program with the goal of implementing current perinatal immunoprophylaxis recommendations throughout the United States. Case managers in this program provide counseling and encourage appropriate testing and vaccination of household contacts. They also coordinate appropriate vaccination and testing of the infant and medical treatment for the mother during pregnancy and postpartum.[28]

Antiviral Therapy

Over the last 10 years, significant progress has been made in the treatment of hepatitis B. In 2012, we have 7 approved drugs for treatment of hepatitis B; 2 interferon-based regimens and 5 nucleoside/nucleotide analogues, as shown in Table 2.

Table 2. Approved drugs for treatment of hepatitis B

Drug	Year Approved
Interferon alpha-2b recombinant	1992
lamivudine	1998
Adefovir	2002
Entecavir	2005
Pegylated interferon alpha-2a	2005
Telbivudine	2006
Tenofovir	2008

All of these drugs are category C except tenofovir and telbivudine (both of which are category B drugs). In general the choice of any medication to be administered during pregnancy must take into consideration the benefit (in this case treatment of active disease or a flare/re-activation of disease causing liver failure, and for preventing maternal to child transmission of hepatitis B) versus the risk (harm to the mother, fetus, and breastfeeding related issues). Currently, no antiviral agent has been approved for use in pregnancy.

In 2003, van Zonneveld, treated 8 highly viremic mothers with 150 mg of lamivudine daily during the last month of pregnancy. All children received passive-active immunization at birth. At 12 month follow-up, 1 out of 8 (12.5%) was still HepBsAg and HBV DNA positive. All the others seroconverted to HepBsAb and maintained seroprotection. In the control group, perinatal transmission occurred in 7 out of 25 (28%).[16] Since then there has been multiple RCT, each with small numbers, evaluating the safety and efficacy of lamivudine treatment in late pregnancy with high HBV DNA levels in preventing mother to child transmission of hepatitis B. A recent meta-analysis suggests that lamivudine treatment in HBV carriers from 28 week of gestation may interrupt mother to child transmission of HBV efficiently, particularly if the viral load is reduced to $<10^6$ copies/mL.[17]

Recently, Han et al conducted a prospective study that evaluated the efficacy of telbivudine for preventing HBV newborn infection in 230 HepBsAg positive pregnant patients with HBV DNA levels of $> 1 \times 10^6$ copies/mL. The study showed that telbivudine plus vaccination is superior to vaccination alone in newborns to prevent HBV transmission (0% vs 8%).[18]

Safety of Antiviral Therapy

All nucleoside/nucleotide analogues are class C except tenofovir and telbivudine (both are class B). This classification is based on the risk of teratogenicity in preclinical evaluation. There are studies of lamivudine in pregnant patients from as early as 2003. Additionally, there are safety data from HIV positive women who have received tenofovir and/or lamivudine for HIV during pregnancy. The Antiretroviral Pregnancy Registry (www.apregistry.com) suggests that tenofovir and lamivudine do not increase the rate of birth defects compared to the national average. Based on the existing evidence, lamivudine, tenofovir and telbivudine could be considered safe to use especially in the third trimester of pregnancy, when the organogenesis of the fetus has been completed. Which drug to use, for how long, potential side effects of the drugs, viral load monitoring, and breastfeeding related issues should be discussed with mother prior starting therapy. Additionally, we should emphasize compliance because of the emergence of resistance or even possibly hepatitis B flare if the patient does not take her medications as directed.

Untreated Pregnant Patients with High Viral Load

In clinical practice, untreated pregnant mothers with viral load > 7log IU/mL will probably require anti-viral therapy to prevent vertical transmission. Tenofovir, telbivudine and lamivudine are eligible choices, but which one to use will depend on affordability, availability, local familiarity with the drug, clinical status of hepatitis B, severity of liver disease, and the need for long term treatment. Patients who are in immunoactive phase of chronic hepatitis B should probably be continued on treatment. It is unclear for how long we should treat patients who are in an immunotolerant phase of hepatitis B. Pregnant women who want to breastfeed after delivery should stop their medication, as breastfeeding is not recommended during anti-viral therapy. Because hepatitis B can flare after treatment discontinuation, these women should be monitored carefully.[19]

Mode of Delivery

Elective cesarean section to prevent perinatal transmission during delivery is appealing. It is thought that perhaps this mode of delivery limits the duration of direct contact between the fetus and the infected secretions or blood in the maternal genital tract. However, there is currently no good quality large RCT to assess the safety and efficacy of elective cesarean section in preventing mother to child HBV transmission. ACOG do not currently endorse this practice.

Breastfeeding

Mothers who are taking nucleoside/nucleotide inhibitors either for treatment of chronic hepatitis B or prevention of vertical transmission should be reminded not to breastfeed. Lamivudine and tenofovir are known to be excreted into breast milk and we do not currently yet have safety data on the effects of these medications on babies.

Studies have documented no difference in rates of perinatal infection between breastfed and formula-fed infants.[20] Thus, HBV infection is not considered a contraindication to breastfeeding for infants who receive HBV and HBIG vaccine.

Summary

Hepatitis B is a major worldwide health problem. It is the leading cause of chronic liver disease, cirrhosis and liver cancer. Immunoprophylaxis has been shown to decrease the rate of maternal to infant transmission of hepatitis B. Every effort should be made to adhere to WHO and CDC guidelines. The emerging data on the impact of mass vaccinations on the decrease in the incidence of cirrhosis and liver cancer in countries like Taiwan are promising.

Hepatitis C

Background

Hepatitis C is a major public health problem in the United States. It is estimated that 170 million people in the world are infected with hepatitis C.[29] It is also the number one cause for liver transplantation in the United States. In the US, the prevalence of HCV infection between the years of 1999 and 2002 was 1.6%, equating to about 4 million people positive for anti-HCV antibody, 80% of whom are estimated to be actively infected. Hepatitis C is the leading cause of liver disease in the US and accounts for the majority of liver cancer in the US.[30] The peak incidence of hepatitis C infection occurred around 1990. Without treatment, the majority of chronically infected hepatitis C patients will progress to cirrhosis over the next 20-30 years. It is estimated that the peak of incidence of cirrhosis from hepatitis C will occur around 2030. In the US, the total medical costs from HCV infection is expected to increase from $30 billion in 2009 to >$85 billion in 2024 (based on treatment with pegylated interferon/ribavirin).[31]

Why Worry about Testing for Hepatitis C?

Approximately 80% of patients who get infected with hepatitis C fail to clear the virus. After 6 months of being infected, if the patient still has detectable virus it is considered chronic hepatitis C. Approximately 20% of these chronic hepatitis C patients will progress to cirrhosis, which will ultimately lead to decompensation (ascites, edema, hepatic encephalopathy, variceal bleeding) or liver cancer.[32] This is in contradistinction to hepatitis B, where if infected as an adult, about 90-95% of patients successfully clear the virus.

High risk patients should be screened for hepatitis C. It is noteworthy that current screening practices fail to identify a large proportion of patients with chronic hepatitis C infection.[33] AASLD recommends that as part of a comprehensive health evaluation, all persons should be screened for behaviors that place them at high risk for HCV infection.[34] Although there is currently no approved medication for the treatment of hepatitis C during pregnancy, newly diagnosed patients should be sent to a liver specialist for further education regarding future options of therapy and preventive health measures to prevent the progression of liver disease.

Who Are High-Risk Patients?

The AASLD 2009 guidelines suggests that the following people are high risk for acquiring hepatitis C infection and should be screened for hepatitis C:[34]

1. Persons who have ever injected illicit drugs, including those who injected only once and do not consider themselves to be drug users,
2. Persons with conditions associated with high prevalence of HCV infection including
   a. Persons with HIV infection,
   b. Persons with hemophilia who received clotting factor concentrates prior to 1987,
   c. Persons who have ever been on hemodialysis, and
   d. Persons with unexplained abnormal LFTs
3. Recipients of transfusions or organ transplants prior to July 1992 including
   a. Persons who were notified that they had received blood from a donor who later tested positive for HCV infection
   b. Persons who received a transfusion of blood or blood products, and
   c. Persons who received an organ transplant
4. Children born to HCV-infected mothers
5. Health care, emergency medical and public safety workers after a needle stick injury or mucosal exposure to HCV-positive blood
6. Current sexual partners of HCV-infected persons

How Do We Make a Diagnosis?

Over the last decade, we have made remarkable improvement in the diagnostic sensitivity of assays to detect HCV RNA. High risk patients should be screened using the latest generation of anti-HCV ELISA testing. A positive test suggests exposure to the virus and should proceed with ordering HCV RNA PCR to determine if the patient has resolved hepatitis C infection or is chronically infected with hepatitis C virus.

What Treatment Options Are Available in 2012?

Treatment of hepatitis C has come a long way since its discovery. For the last 10 years, the main therapy for hepatitis C revolved around interferon and ribavirin. In 2011, we were excited to see the approval of two new protease inhibitors (telaprevir and boceprevir) for use together with the standard pegylated interferon and ribavirin for genotype 1 patients. The new triple therapy regimen significantly improves the cure rate for hepatitis C genotype 1 patients from an average of 45 to about 70%.

Depending on the genotype, severity of live disease, viral kinetics, side effects to treatment etc., the treatment duration may be anywhere from 6 months to 12 months.

Hepatitis C Treatment and Pregnancy

Treatment for hepatitis C is contraindicated during pregnancy. Pegylated interferon is a Pregnancy Category C drug while ribavirin, boceprevir and telaprevir are Pregnancy Category X drugs. However, it is important to remember to refer patients back to see their liver specialist to assess for eligibility for treatment at some point in time after delivery. In general, a good time would be 6-12 months after delivery. Hepatitis C treatment requires a significant commitment because the treatment process requires frequent blood work, frequent doctor visits, has multiple side-effects, and usually causes fatigue. Additionally, mothers with postpartum depression are not treatment candidates until their depression is under control.

Patients who have active hepatitis C infection should be educated about the mode of transmission, to stop smoking, abstain from alcohol and to get vaccinated for hepatitis A and B.

In general, all patients who are currently undergoing treatment should use at least two forms of contraception to prevent accidental pregnancy during treatment. Additionally, they should have monthly pregnancy tests. Patients should not get pregnant for at least 6 months after they have stopped hepatitis C treatment.

Epidemiology

Seroprevalence of HCV in pregnant women is thought be anywhere from 0.15 to 2.4% in the US and European countries but estimated to be as high as 8.6% in Egypt.[35-38] The prevalence in children ranges from 0.05 to 0.36% in the US and Europe but much higher from 1.8 to 5% in developing countries.[39,40]

In the developed countries including the US, vertical transmission is the major route for HCV infection. In the developing countries, contaminated blood products and parenteral exposure continue to be the major causes of HCV infection.[41]

Natural History of HCV in Pregnancy

In the great majority of patients, pregnancy does not seem to affect the clinical course of acute or chronic hepatitis C. HCV cirrhotics are usually not fertile enough to conceive, but in the event that they do, they would require special care, as there is a possibility of decompensation due to worsening of portal hypertension and bleeding during delivery due to coagulopathy.[42,43]

Biochemically, HCV RNA tends to increase during the second and third trimester and decrease after delivery. But in some women, HCV RNA has been reported to rise toward the end of pregnancy.[46] Transaminases tend to improve during pregnancy.[43,44] Decrease in liver enzymes is thought to be related to hemodilution in pregnancy. Alternatively, pregnancy is also thought be induce a relative state of immunosuppression towards the third trimester to prevent rejection of the fetus, hence the higher level of viral load and decrease in liver enzymes. It is interesting to note that viral load tend to decrease after delivery, perhaps because of the reconstitution of the immune system after delivery.[45]

Obstetrically, there have been varied reports on the complications in HCV-infected women. In an Irish study of women inadvertently infected with HCV-contaminated immunoglobulin anti-D, there was no increase in fetal or obstetrical complications.[47] Pergam et al, in a population based cohort study found that infants of HCV-infected mothers were more likely to be low birth weight, to be small for gestational age, to need assisted ventilation and to require neonatal intensive care unit admission.[48]

Perinatal Transmission

As there are no currently approved vaccines for hepatitis C and no approved treatment for pregnant women, many studies have focused on perinatal transmission, looking for ways to decrease the risk of transmission from mother to infant. Fortunately, in the US, the risk of transmission remains small for women with hepatitis C at approximately 4%. In a systematic review of perinatal transmission rates from a large number of studies published between 1992-2000, the weighted rate of perinatal transmission of HCV was 1.7% when the mother was anti-HCV positive irrespective of HCV RNA, 4.3% when the mother was positive for HCV RNA and 19.4% when the mother was co-infected with HIV.

Timing of Infection

It is unclear if transmission of HCV occurs intrauterine or intrapartum. In the majority of infants who are infected with hepatitis C, HCV RNA reaches detectable levels a few weeks after birth suggesting that HCV is acquired intrapartum or late intrauterine.[51] However, Mok et al also showed that in some infants, HCV RNA is detectable within the first 24 hours of life, suggesting that HCV was acquired early during intrauterine life.[58]

Testing for Hepatitis C in Infants

The NIH consensus statement in 2002 recommends the following testing for the diagnosis for perinatal hepatitis C infection:

1. Infants born to HCV-positive mothers should be tested for HCV infection by HCV RNA tests on two occasions between ages of 2 and 6 months and/or
2. Have tests for anti-HCV after 15 months of age.

All infants born to chronic hepatitis C mothers will test positive for anti-HCV antibody regardless of whether they have been infected or not (via transplacental transfer of IgG antibody). Because the passively transferred antibodies can persists up to 18 months of age, the AASLD 2009 Hepatitis C Guidelines recommends testing for anti-HCV antibody at 18 months of age or older.[49]

The sensitivity of HCV RNA testing is low at birth and improves by 6 months of age. Polywka et al showed that the sensitivity was only 22% at birth but increased to 85% by 6 months.[50] A similar study done by Gibb et al showed that PCR sensitivity was only 22% at first month and rose rapidly thereafter to about 97%.[51]

Infants born to HCV RNA negative mothers should have LFTs and anti-HCV tested at 18 months. If negative, no further follow-up is needed.

Infants born to HCV RNA positive mothers should have LFTs and anti-HCV tested at 18 months. If LFTs are normal, HCV RNA negative, but anti-HCV antibody remains positive, this suggests evidence of past infection.

Factors Increasing the Risk of Perinatal HCV Infection

HIV Co-Infection

Maternal HCV/HIV co-infection has been consistently shown to increase the risk of HCV transmission to infants. The risk is increased approximately 3-4 folds depending on the studies.[52,53] The mechanism of increased risk is unclear. It has been suggested that HIV causes high level of maternal HCV viremia by immunosuppression, thus facilitating transmission.[54] Blackard also demonstrated that HIV infection facilitates HCV entry and replication in blood mononuclear cells, which is also thought to facilitate transmission. It is not known if treatment of HIV decreases rate of transmission of hepatitis C to infant.

Intravenous Drug Use

Several studies showed an increase risk of perinatal transmission of HCV in pregnant women actively using intravenous drugs.[56,57] Infection of peripheral blood mononuclear cells with HCV occurs more frequently in IV drug users which may favor the transmission of HCV. However, the exact mechanism remains unknown.

Obstetrical Factors

A few studies have found that premature rupture of membrane more than 6 hours before delivery is associated with increased risk of perinatal transmission of HCV.[59,60] Steininger et al reported increased risk of perinatal transmission of HCV by intrapartum infantile exposure to maternal blood, such as via fetal scalp electrode.[61]

Minola et al suggests that amniocentesis could potentially be a risk factor for mother to infant transmission of HCV, although there is scarcity of data in this area.[62]

Factors Not Associated with Increased Risk of Perinatal HCV Infection

Mode of Delivery

Multiple observational studies showed that cesarean section or vaginal delivery did not make a difference in perinatal transmission rate. Currently there are no randomized control trials aimed at evaluating whether cesarean section rather than vaginal delivery reduces HCV perinatal transmission. In a 2009 meta-analysis, Tabatabaie et al identified 8 studies involving 641 unique mother-infant pairs. The study suggests that cesarean section does not decrease perinatal HCV transmission from HCV RNA+/HIV- mothers to infants.[63]

Breastfeeding in Hepatitis C

Hepatitis C virus has been found in low levels in breast milk of HCV RNA positive mothers.[64] Most studies indicate that even though theoretical transmission may be possible, the viral load in breast milk is extremely low and likely becomes inactivated in the digestive tract of the infant.[65] The European Paediatric Hepatitis C Virus Network showed no difference in the infection rates in breastfed versus formula-fed infants in a study carried out on 1758 infants.[66] The CDC recommends that it is safe for mothers infected with HCV to breastfeed unless she has cracked or bleeding nipples.

Summary

Although we do not currently have any approved treatment for hepatitis C during pregnancy, identifying high risk patients and diagnosing hepatitis C is important. They should be referred to a liver specialist to discuss possible options for treatment once the pregnancy is over. There is currently no available therapy to prevent vertical transmission of HCV.

Other Less Common Causes of Viral Hepatitis in Pregnancy

Herpes Simplex Virus (HSV) Hepatitis

Herpes simplex virus hepatitis is a rare disorder but commonly fatal condition. The mortality in untreated patients is >80%.[67] Patients die from fulminant acute liver failure. It is important to think about this condition in the appropriate setting, because treatment with acyclovir can be life-saving to the mother and fetus. Additionally, HSV hepatitis needs to be differentiated from pregnancy-associated acute liver disease (PAALD) (e.g., acute fatty liver of pregnancy, HELLP syndrome, preeclampsia-associated liver disease) because the definitive treatment of PAALD is delivery of the fetus.

Disseminated herpes virus infection is rare in the immunocompetent adult. It has been reported in immunocompromised adults including renal transplant patients, bone marrow transplant recipients and those receiving steroid therapy.[68] Pregnant women, particularly in their third trimester, are also susceptible. A recent review found that 32 of the 137 reported cases of HSV hepatitis occurred in pregnant women.[68]

HSV hepatitis may result from HSV serotype 1 and serotype 2, as well as both primary and latent infections.[69] Clinically, most patients present with nonspecific flu-like illness prodrome with fever (98%), herpetic lesion (56%), mean ALT or AST 4927 ± 4099, mean peak bilirubin 6 ± 8.1, leucopenia (71.4%), thrombocytopenia (93.6%), coagulopathy (83.7%), acute renal failure (65.3%), hepatic encephalopathy (80%).[68] Liver transaminases in HSV hepatitis generally increase 100-1000-fold above normal.[68] In the absence of HSV serology and PCR markers, it can be difficult to distinguish this condition from PAALD as both may present with abnormal LFTs, jaundice, low platelets, anemia and change in mental status. Also, the characteristic rash in herpes is absent in almost half of the patients, making this a less useful sign to look for when trying to make the diagnosis. In fact, more than 50% of patients with HSV hepatitis are only diagnosed at autopsy, suggesting that physicians are not considering this diagnosis or are diagnosing them too late.[68]

Reports on the CT-scan of the liver in pregnancy is limited. Mortele et al described CT features of HSV hepatitis with multiple, hypodense, subcentimeter nodules scattered throughout the liver.[70] Similar radiographic findings have been described in a case of fulminant herpes simplex hepatitis in a case of Crohn's disease.[71] In this case, the multiple hypodense lesions in the liver corresponded to focal areas of liver necrosis.

Liver biopsy is invasive but remains the gold standard for diagnosis of this condition. All patients suspected of having HSV hepatitis should have IgM and IgG serology for HSV drawn. HSV PCR is not routinely performed, although this is thought to be the most accurate and least invasive method to diagnose HSV hepatitis. The drawback of serology is getting the results back in time to allow a treatment decision to be made. In the absence of the availability of serology for HSV, liver biopsy may be a useful tool to make the diagnosis of HSV hepatitis. Suspicion for this condition should be noted in the pathology request, so that the lab may perform the appropriate immunostaining. Like serology, liver biopsy results will take time. Because of the high rate of morbidity and mortality of HSV hepatitis in pregnancy, patients should be treated empirically with acyclovir until serology or biopsy results come back.

There are no randomized control trial for the treatment of HSV hepatitis in pregnant patients. But retrospective data and case reports data support acyclovir use.[72,73] Successful liver transplantation for fulminant HSV hepatitis has also been reported in the literature.[74,75]

Hepatitis E Virus Infection

Hepatitis E, caused by hepatitis E virus (HEV), it was first recognized during an epidemic of hepatitis which occurred in the Kashmir valley in 1978.[76] Hepatitis E was cloned and sequenced in 1990.[77] HEV is an enterically transmitted virus particle of 27-34nm diameter whose genome consists of a poly-adenylated single stranded RNA of 7.2kb length.[78] It is included in this chapter because of its morbidity and propensity to progress to fulminant hepatic failure in infected pregnant women in certain countries, particularly India. Clinically apparent acute hepatitis E is rare in the US.

Because of its enteric mode of transmission, it is the most common cause of acute icteric and anicteric hepatitis in areas of poor sanitation.[79] Multiple epidemics of hepatitis E have been reported in India, China, Southeast and Central Asia, Northern and Western Africa.[80] The sources of these epidemics have been traced to contaminated drinking water. However, transmission through contaminated food, infected blood products, and maternal-fetal transmission have also been documented in hyper-endemic regions like the Indian subcontinent.[79]

Acute hepatitis E in nonpregnant patients and men is generally a self-limiting disease and resolves within 4-6 weeks with case fatality rates in hospital varying between 0.5 to 4%.[79] It has long been observed that hepatitis E has a high incidence and a more severe course in pregnant women in some geographical areas where HEV is endemic, such as Northern India.[81,82] What is not clear is why in other HEV-endemic countries like Egypt, pregnant patients infected with HEV have little or no morbidity.[83] Reports from the US and Europe, where HEV is not endemic suggest that severity of hepatitis E in pregnant women is not different from that in nonpregnant women. In a review paper, Navaneethan et al suggests that severe liver injury because of HEV infection during pregnancy may be related to several possible causes, such as immune and humoral factors occurring during pregnancy, genetic and environmental factors.[84] In a letter to the editor, another group of hepatitis E investigators suggests that the differences in severity of hepatitis E in pregnant women may be related different genotypes of HEV in different geographical regions.[85]

In 2009, Khuroo et al have studied the clinical course and duration of viremia in vertically transmitted HEV infection in babies born to HEV-infected mothers. 19 babies were studied. Babies were studied at birth and on a monthly basis to evaluate clinical profile, pattern of antibody response and duration of viremia in those infected with HEV. 78.9% had evidence of vertically transmitted HEV infection at birth. Of the HEV infected babies, 40% died. All surviving babies had self-limiting disease and none had prolonged viremia. The authors concluded that HEV infection is commonly transmitted from mother to fetus and causes high neonatal mortality rate. HEV infection in survivors is self limiting with short lasting viremia.[86]

This infection should be in the differential diagnosis of any pregnant patients with hepatitis in endemic areas or any pregnant patient from a non-endemic country who had recently travelled to a HEV-endemic region. Treatment remains supportive. Recent awareness of chronic HEV infection in patients and possible progression of liver disease has led to attempts to treat hepatitis E with pegylated interferon and/or ribavirin.[87,88] Ribavirin is teratogenic and its use is not indicated for pregnant patients. In the last decade two vaccines for hepatitis E have been developed. Perhaps in the future, these may be useful for persons who are at a high risk of severe disease following HEV infection such as pregnant patients, patients with chronic liver diseases and immunosuppressed patients at risk of chronic HEV infection.[89] For now, travelers to endemic areas should be cautioned to avoid drinking water of unknown purity, uncooked shellfish and uncooked fruits or vegetables. The role of vaccines for hepatitis E remains unclear.[89]

Hepatitis A

Hepatitis A is a 27-32 nm RNA virus that is transmitted mainly by the oral-fecal route, as well as through contaminated water and food. Symptoms and signs of acute infection usually occur within 2 to 4 weeks of exposure.

If a pregnant woman becomes infected, the baby is generally not affected. Intrauterine transmission of hepatitis is rare.[90,91] A few cases of vertical transmission of hepatitis A have been reported in the literature.[92] It is generally accepted hepatitis A infection is not associated with severe outcome or complications during pregnancy as illustrated in a case report by Mette et al. However, Elinav et al have reported that acute hepatitis A infection in pregnancy is associated with high rates of gestational complications and preterm labor. In this study, the authors evaluated 79, 458 pregnant females over a 25 year period, and found 13 cases of HAV infections. 9 of 13 (69%) developed gestational complications, including premature contractions (n = 4), placental separation (n = 2), premature rupture of membranes (n = 2) and vaginal bleeding (n = 1).[93]

Women susceptible to hepatitis A virus should be vaccinated for hepatitis A because of the associated risk of preterm delivery. A theoretical risk to fetus cannot be excluded, but as with other inactivated vaccines, it is very likely to be low.

Hepatitis A infection is not a contraindication for breastfeeding. As there is no evidence of transmission via breast milk, women are encouraged to breastfeed.

Conclusion

Table 3 provides a summary of various types of viral hepatitis regarding effect on pregnancy, perinatal transmission rate, prevention of perinatal transmission and breastfeeding.

Table 3. Summary of characteristics in various types of viral hepatitis

Hepatitis	Effect on Pregnancy	Perinatal Transmission Rate	Prevention of Perinatal Transmission	Breastfeeding
A	Associated with preterm labor and maternal complications. Does not appear to increase risk of congenital malformation, stillbirths, IUGR, spontaneous abortions	Rare cases	Vaccinate mother who are high risk of exposure If exposed, immunoglobulin	OK
B	Associated with preterm labor. Associated with LBW	Acute hepatitis in 1st trimester rare. Acute hepatitis in 3rd trimester risk of maternal infant transmission rate increases 90% Chronic HepBeAg+ 10% Chronic HepBeAg–	Vaccinate mother who are high risk of exposure HBV Vaccines and HBIG Consider antiviral therapy	OK as long as newborn got immunoprophylaxis. No if mother on antivirals
C	Little data on acute hepatitis C. For chronic cases, no adverse effect on pregnancy. Has been associated with LBW, SGA, needing NICU admissions.	HCV ~ 5% HCV—HIV ~ as high as 36%	No known therapy	OK provided nipples are not cracked or bleeding
E	May develop fulminant hepatitis when acquired in 3rd trimester. Associated with worse obstetric outcome. High fetal mortality rate.	One study suggests high rate of 79%, with 40% mortality	No known therapy	OK

References

1. Lavanchy D. Hepatitis B virus epidemiology, disease burden, treatment, and current and emerging prevention and control measures. J Viral Hepat 2004; 11(2):97-107.
2. Wasley A, Grytdal S, Gallagher K. Surveillance for acute viral hepatitis—United States 2006. MMWR Surveill Summ 2008; 57:1-24.
3. Cotler SJ, Dhamija MK, Luc BJ et al. The prevalence and clinical correlates of elevated ALT levels in an urban Chinatown community. J Viral Hepat 2010; 17:148-152.
4. Lok AS, McMahon BJ. Chronic hepatitis B: update 2009. Hepatology 2009; 50:661-662.
5. Loo NM, Wagie LA, Rahman ML et al. Effectiveness of Hepatitis B screening in primary care setting. Hepatology 2010; 52 (Issue 1): 627A-724A.
6. World Health Organization. Hepatitis B vaccines. Wkly Epidemiolo Rec 2009; 84:405-419.
7. Wong S, Chan LY, Yu V et al. Hepatitis B carrier and perinatal outcome in singleton pregnancy. Am J Perinatal 1999; 16:485-488.
8. Safir A, Levy A, Sikuler E et al. Maternal hepatitis B virus or hepatitis C virus carrier status as an independent risk factor for advserse perinatal outcome . Liver Int 2010; 30(5):765-770.
9. Elefsiniotis IS, Papadakis M, Vlachos G et al. Presence of HBV-DNA in Cord Blood is associated with spontaneous preterm birth in pregnant women with HBeAg negative chronic hepatitis B virus infection. Intervirology 2011; 54(5):300-304.
10. Peters MG. Special populations with hepatitis B virus infection. Hepatology 2009.
11. Nguyen G, Garcia RT, Nguyen N et al. Clinical course of hepatitis B virus infection during pregnancy.
12. Sweet RL. Hepatitis B infection in pregnancy. Obstet Gynecol Rep 1990; 2:128.
13. Ter Borg MJ, Leemans WF, de Man RA. Exacerbation of chronic hepatitis B infection after delivery. J Viral Hepat 2008; 15(1):37-41.
14. Mast EE, Margolis HS, Fiore AE et al. A comprehensive immunization strategy to eliminate transmission of hepatitis B virus infection in the United States: recommendations of the Advisory Committee on Immunization Practices (ACP) part 1: immunization of infants, children, and adolescents. MMWR Recomm Rep 2005; 54(RR-16):1-31.
15. Wiseman E, Fraser MA, Holden S et al. Perinatal transmission of hepatitis B virus: an Australian experience. Med J Aust 2009; 190:489-492.
16. Von Zonneveld M, van Nunen AB, Niesters HG et al. Lamivudine treatment during pregnancy to prevent perinatal transmission of hepatitis B virus infection. J Viral Hep 2003; 10(4):294-297.
17. Han L, Zhang HW, Xie JX et al. A meta-analysis of lamivudine for interruption of mother to child transmission of hepatitis B virus. World J Gastroenterol 2011; 17(38):4321-4333.
18. Han GR, Cao MK, Zhao W et al. A Prospective and open-label study for the efficacy and safety of telbivudine in pregnancy for the prevention of perinatal transmission of hepatitis B virus infection. J Hepatol 2011; 55(6):1215-1221.
19. Pol S, Corouge M, Fontaine H. Hepatitis B virus infection and pregnancy. Clin Res Hepatol Gastroenterol 2011; 35(10):618-622.
20. Beasley RP, Stevens CE, Shiao IS et al. Evidence against breastfeeding as a mechanism for vertical transmission of hepatitis B. Lancet 1975; 2:740-741.
21. Hammitt LL, Hennessy TW, Fiore AT et al. Hepatitis B immunity in children vaccinated with recombinant hepatitis B vaccines beginning at birth: a follow-up study at 15 years. Vaccine 2007; 25:6958-6964.
22. Handler SC, Francis DP, Maynard JE et al. Long term immunogenicity and efficacy of hepatitis B vaccine in homosexual men. N Engl J Med 1986; 315:209-214.
23. Sheffield JS, Hickman A, Tang J et al. Efficacy of an accelerated hepatitis B vaccination program during pregnancy. Obstet Gynecol 2011; 117(5):1130-1135.
24. Roznovsky L, Orsagova I, Kloudova A et al. Long term protection against hepatitis B after newborn vaccination: 20 year follow up. Infection 2010; 38(5):395-400.
25. Beasley RP. Rocks along the road to the control of HBV and HCC. Ann Epidemiol 2009; 19(4):231-234.
26. Ni YH, Chen DS. Hepatitis B vaccination in children: the Taiwan experience. Pathol Biol (Paris) 2010; 58(4):296-300.

27. Tajiri H, Tanaka H, Brooks S et al. Reduction of hepatocellular carcinoma in childhood after introduction of selective vaccination against hepatitis B virus for infants born to HBV carrier mothers. Cancer Causes Control 2011; 22(3):523-527.
28. Degli Esposti S, Shah D. Hepatitis B in pregnancy: challenges and treatment. Gastroenterol Clin North Am 2011; 40(2):355-372.
29. NIH Consensus Statement on management of hepatitis C: 2002. NIH Consens State Sci Statements 2002; 19(3):1-46.
30. Armstrong GL, Wasley A, Simard EP et al. The prevalence of hepatitis C in the United States, 1999 through 2002. Ann Intern Med 2006; 144:705-714.
31. Milliman Inc. Consequences of hepatitis C: costs of a baby boomer epidemic of liver disease 2009.
32. Alberti A, Vario A, Ferrari A et al. Review article: chronic hepatitis C natural history and cofactors. Aliment Pharmacol Ther 2005; 22:74-78.
33. Kim WR. The burden of hepatitis C in the United States. Hepatology 2002; 36:S30-S34.
34. Ghany MG, Strader DB, Thomas DL et al. Diagnosis, management and treatment of hepatitis C: an update. Hepatology 2009; 49:1335-1374.
35. Costa ZB, Machado BD, Avelino MM et al. Prevalence and risk factors for hepatitis C and HIV-1 infections among pregnant women in Central Brazil. BMC Infect Dis 2009; 9:116.
36. Alter MJ, Kruszon-Moran D, Nainan OV et al. The prevance of hepatitis C virus infection in the United States, 1988 through 1994. N Engl J Med 1999; 341(8):556-562.
37. Conte D, Fraquelli M, Prati D et al. Prevalence and clinical course of chronic hepatitis C virus (HCV) infection and rate of HCV vertical transmission in a cohort of 15,250 pregnant women. Hepatology 2000; 31(3):751-755.
38. AbdulQawi K, Youssef A, Metwally MA et al. Prospective study of prevalence and risk factors for hepatitis C in pregnant Egyptian women and its transmission to their infants. Croat Med K 2010; 51(3):219-228.
39. Jafri W, Jafri N, Yakoob J et al. Hepatitis B and C: prevalence and risk factors associated with seropositivity among children in Karachi, Pakistan. BMC Infect Dis 2006; 6:101.
40. El-Kamary SS, Serwint JR, Joffe A et al. Prevalence of hepatitis C virus infection in urban children. J Pediatr 2003; 143(1):54-59.
41. Akhtar S, Moatter T. Hepatitis C virus infection in polytransfused thalassemic children in Pakistan. Indian Pediatr 2004; 41(10):1072-1073.
42. Britton RC. Pregnancy and esophageal varices. Am J Surg 1982; 143(4):412-415.
43. Cheng Y. Pregnancy in liver cirrhosis and/or portal hypertension. Am J Obst Gynecol 1977; 128:812.
44. Gervais A, Bacq Y, Bernuau J et al. Decrease in serum ALT and increase in serum HCV RNA during pregnancy in women with chronic hepatitis C. J Hepatolo 2000; 32(2):293-299.
45. Lin HH, Kao JH. Hepatitis C virus load during pregnancy and puerperium. BJOG 2000; 107(12):1503-1506.
46. Paternoster DM, Santarossa C, Grella P et al. Viral load in HCV RNA-positive pregnant women. Am J Gastroenterol 2001; 96:2751.
47. Jabeen T, Cannon B, Hogan J et al. Pregnancy and pregnancy outcome in hepatitis C type 1b. QJM 2000; 93(9):597-601.
48. Pergam SA, Wang CC, Gardella CM et al. Pregnancy complications associated with hepatitis C: data from a 2003-2005 Washington state birth cohort. Am J Obstet Gynecol 2008; 199(1): e1-e9.
49. Mast EE, Hwang LY, Seto DS et al. Risk factors for perinatal transmission of hepatitis C virus (HCV) and the natural history of HCV infection acquired in infancy. J Infect Dis 2005; 192:1880-1889.
50. Polywka S, Pembrey L, Tovo PA. Accuracy of HCV-RNA PCR tests for diagnosis or exclusion of vertically acquired HCV infection. J Med Virol 2006; 78(2):305-310.
51. Gibb DM, Goodall RL, Dunn DT et al. Mother-to-child transmission of hepatitis C virus: evidence for preventable peripartum transmission. Lancet 2000; 356:904-907.
52. Polis CB, Shah SN, Johnson KE et al. Impact of maternal HIV co-infection on the vertical transmission of hepatitis C virus: a metaanalysis. Clin Infec Dis 2007; 44(8):1123-1131.

53. Marine—Barjoan E, Berrebi A, Giordanengo V et al. HCV/HIV co-infection, HCV viral load and mode of delivery: risk factors for mother to child transmission of hepatitis C virus. AIDS 2007; 21(13):1811-1815.
54. Thomas DL, Villano SA, Riester KA et al. Perinatal transmission of hepatitis C from human immunodeficiency virus type 1 infected mothers. J Infect Dis 1998; 177:480-488.
55. Blackard JT, Smeaton L, Hiasa Y et al. Detection of hepatitis C virus in serum and peripheral blood mononuclear cells from HCV monoinfected and HIV/HCV co-infected persons. J Inct Dis 2005; 192:258-265.
56. Resti M, Azzari C, Moriondo M et al. Injection drug use facilitates hepatitis C virus infection of peripheral blood mononuclear cells. Clin Infect Dis 2000; 35:236-239.
57. Azzari C, Moriondo M, Indolfi G et al. Higher risk of hepatitis C virus perinatal transmission from drug user mothers is mediated by peripheral blood mononuclear cell infection. J Med Virol 2008; 80:65-71.
58. Mok J, Pembrey L, Tovo PA et al. European Paedeatric Hepatitis C Virus Network 2005. When does mother to child transmission of hepatitis C virus occur? Arch Dis Child Fetal Neonatal Ed 90:156-160.
59. Mast EE, Hwang LY, Seto DS et al. Risk factors for perinatal transmission of hepatitis C virus (HCV) and the natural history of HCV infection acquired in infancy. J Infect Dis 2005; 192(11):1880-1889.
60. Thomas SL, Newell ML, Peckham CS et al. A review of hepatitis C virus (HCV) vertical transmission: risk of transmission to infants born to mothers with and without HCV viraemia or human immunodeficiency virus infection. Int J Epidemiol 198; 27(1):108-117.
61. Steininger C, Kundi M, Jatzko G et al. Increased risk of-mother-to infant transmission of hepatitis C virus by intrapartum infantile exposure to maternal blood. J Infect Dis 2003; 187:345-351.
62. Minola E, Maccabruni A, Pacati L. Amniocentesis as a possible risk factor for mother to infant transmission of hepatitis C virus. Hepatology 2011; 33:1341-1342.
63. Ghamar Chehreh ME, Tabatabaei SV, Khazanehdari S et al. Effect of cesarian section on the risk of perinatal transmission of hepatitis C virus from HCV-RNA+/HIV– mothers: a meta-analysis. Arch Gynecol Obstet 2011; 283(2):255-260.
64. Ruiz-Extremera A, SalmeronJ, Torres C et al. Follow up of transmission of hepatitis C to babies of human immunodeficiency virus negative women: the role of breast feeding in transmission. Pediatr Infect Dis J 2000; 19:511-516.
65. Yeung LT, King SM, Roberts EA. Mother to infant transmission of hepatitis C virus. Hepatology 2001; 34(2):223-229.
66. European Paediatric Hepatitis C Virus Network. A significant sex—but not elective cesarean section—effect on mother to child transmission of hepatitis C virus infection. J Infect Dis 2005; 192(11):1872-1879.
67. Kaufman B, Ghandi SA, Louise E et al. Herpes simplex virus hepatitis: case report and review. Clin Infect Dis 1997; 24:334-338.
68. Norvell JP, Blei AT, Jovanovic BD et al. Herpes simplex virus hepatitis: an analysis of the published literature and institutional cases. Liver Transpl 2007; 13:1428-1434.
69. Kang AH, Graves CR. Herpes simplex hepatitis in pregnancy: a case report and review of the literature. Obstet Gynecol Surv 1999; 54:463-468.
70. Mortele KJ, Barish MA, Yucel KE. Fulminant herpes hepatitis in an immunocompetent pregnant woman: CT imaging features. Abdom imaging 2004; 29(6):682-684.
71. Wolfsen HC, Bolen JW, Bowen JL et al. Fulminant herpes hepatitis mimicking hepatic abscess. J Clin Gastroenterol 1993; 16:61-64.
72. Peters DJ, Greene WH, Ruggiero F et al. Herpes simplex-induced hepatitis in adults: a call for empiric therapy. Dig Dis Sci 2000; 45(2):2399-2404.
73. Allen RH, Tuomala RE. Herpes simplex virus hepatitis causing acute liver dysfunction and thrombocytopenia in pregnancy. Obstet Gynecol 2005; 106(5 Pt 2):1187-1189.
74. Thurman RH, Konig K, Watkins A et al. Fulminant herpes simplex virus hepatic failure in pregnancy requiring liver transplantation. Aust N Z J Obstet Gynaecol. 2010; 50(5):492-494.

75. Halliday J, Lokan J, Angus PW et al. A case of fulminant hepatic failure in pregnancy. Hepatology 2010; 51(1):341-342.
76. Khuroo MS. Study of an epidemic of non-A, non-B hepatitis: possibility of another human hepatitis virus distinct from posttransfusion non-A, non-B type. Am J Med 1980; 68:818-823.
77. Reyes GR, Purdy MA, Kim J et al. Isolation of cDNA from the virus responsible for enterically transmitted non-A, non-B hepatitis. Science 1990; 247:1335-1339.
78. Balayan MS, Andjaparidze AG, Savinskaya SS et al. Evidence for a virus in non-A, non-B hepatitis transmitted via fecal-oral route. Intervirology 1983; 20:23-31.
79. Acharya SK, Panda SK. Hepatitis E Virus: epidemiology, diagnosis, pathology and prevention. Trop Gastroenterol 2006; 27:63-68.
80. Aggarwal R, Naik S. Epidemiology of hepatitis E: current status. J Gastroenterolo Hepatol 2009; 24:1484-1493.
81. Kumar A, Beniwal M, Kar P et al. Hepatitis E in pregnancy. Int J Gynaecol Obstet 2004; 85:240-244.
82. Patra S, Kumar A, Trivedi SS et al. Maternal and fetal outcomes in pregnant women with acute hepatitis E virus infection. Ann Intern Med 2007; 147:28-33.
83. Stoszek SK, Abdel-Hamid M, Saleh DA et al. High prevalence of hepatitis E antibodies in pregnant Egyptian women. Tras R Soc Trop Med Hyg 2006; 100:95-101.
84. Navaneethan U, Mohajer MA, Shata MT. Hepatitis E and pregnancy: understanding the pathogenesis. Liver Int 2008; 28(9):1190-1199.
85. Renou C, Pariente A, Nicand E et al. Pathogenesis of hepatitis E in pregnancy. Liver Int 2008; 28:1465.
86. Khuroo MS, Kamili S, Khuroo MS. Clinical course and duration of viremia in vertically transmitted hepatitis E virus infection in babies born to HEV-infected mothers. J Viral Hepat 2009; 16:519-523.
87. Mallet V, Nicand E, Sultanik P et al. Brief communication: case reports of ribavirin treatment for chronic hepatitis E. Ann Intern Med 2010; 153:85-89.
88. Kamar N, Rostaing L, Abravanel F et al. Pegylated interferon-alpha for treating chronic hepatitis E virus infection after liver transplantation. Clin Infec Dis 2010; 50:e30-e33.
89. Aggarwal R. Hepatitis E: historical, contemporary and future perspectives. J Gastroenterol Hepatol 2011; 26 Supple 1:72-82.
90. Motte A, Blanc J, Minodier P et al. Acute hepatitis A in a pregnant woman at delivery. Int J Infect Dis 2009; 13(2):e49-e51.
91. Selander B, Blackberg J, Widell A et al. No evidence of intrauterine transmission of hepatitis A virus from a mother to a premature infant. Acta Paediatr 2009; 98(10):1603-1606.
92. Renge R, Danie VS, Chitambar SD et al. Vertical transmission of Hepatitis A. Indian J Pediatr 2002; 69(6):535-536.
93. Elinav E, Ben-Dov IZ, Shapira Y et al. Acute hepatitis A infection in pregnancy is associated with high rates of gestational complications and preterm labor. Gastroenterology 2006; 130(4):1129-1134.

Section III
Liver Disease Unique to Pregnancy

CHAPTER 4

Hyperemesis Gravidarum and Maternal Liver Disease

William M. Outlaw, Jamal A. Ibdah and Kenneth L. Koch*

Abstract

Hyperemesis gravidarum (HG) is the most severe form of illness within the spectrum of nausea and vomiting of pregnancy (NVP). HG affects millions of pregnant women annually, imparts significant cost to society, and places mother and her unborn child at risk. The nausea and vomiting characteristic of HG presents early in pregnancy and usually terminates by the 20th gestational week. Liver disease, usually consisting of mild serum transaminase elevation, occurs in almost 50% of patients with HG. While multiple risk factors have been proposed, the etiology of HG and the maternal liver disease seen in HG remains unclear. Current investigations underscore the potential roles of starvation injury, placental release of inflammatory cytokines, and impairment of fatty acid oxidation in the pathogenesis of the maternal liver disease observed in HG.

> *"Let me speak the plain truth—my sufferings are very great—my nights indescribable —sickness with scarce a reprieve—I strain until what I vomit is mixed with blood."*
>
> —Charlotte Bronte, author of Jane Eyre, purported to have died from hyperemesis gravidarum.

NVP is a common entity which affects up to 90% of pregnancies.[1] The breadth of disease in NVP is considerable. Pathology can range from mild nausea to HG which is marked by unrelenting nausea and vomiting causing dehydration, metabolic derangements, and nutritional deficiencies. The spectrum of NVP is illustrated in Figure 1.

Nausea and Vomiting of Pregnancy

NVP affects the great majority of pregnancies. Over 4 million women suffer with NVP each year and the condition results annually in approximately 8.6 million lost hours of paid employment and 5.8 million lost hours of household work.[2] NVP typically consists of mild nausea and vomiting presenting during the first trimester of pregnancy. Physical examination and laboratory findings in most patients with NVP are within normal limits. NVP is more common in young, nulliparous, and obese women while maternal age greater than 35 years and cigarette smoking may confer protective effects.[3] NVP is frequently seen in Western countries and is relatively rare in African and Asian populations.

*Corresponding Author: Kenneth L. Koch—Section on Gastroenterology, Department of Internal Medicine, Wake Forest University School of Medicine, Winston-Salem, North Carolina, 27157 USA. Email: kkoch@wfubmc.edu

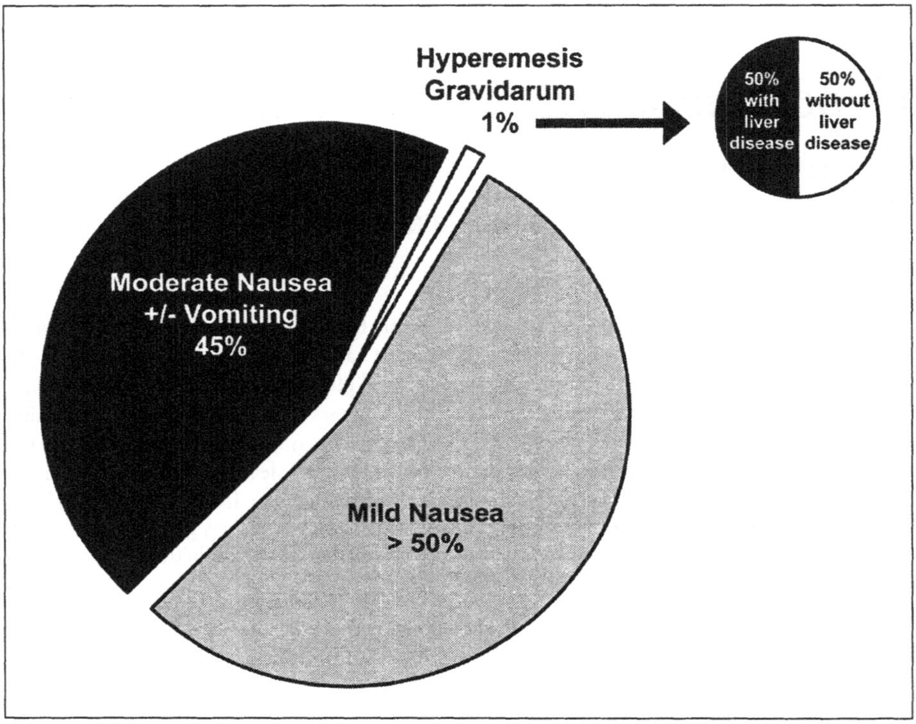

Figure 1. Spectrum of nausea and vomiting of pregnancy. Over 50% of patients with NVP experience mild nausea alone, while nearly 45% have moderate nausea with or without vomiting. While HG represents the most severe form of pathology within NVP, it comprises no more than 1% of all cases of NVP. Liver disease, typically manifesting as mild transaminase elevation, occurs in approximately 50% of patients with HG but is unusual in the other forms of NVP.

NVP is a diagnosis of exclusion and the differential diagnosis of nausea in pregnancy is vast. Mechanical obstruction, peptic and biliary disease, hormonal and metabolic imbalances, medication effects, and central nervous systems disorders must all be ruled out before one makes the diagnosis. Once the diagnosis of NVP is certain, supportive care including dietary modifications with small and frequent meals abundant in protein as well as pharmacologic therapy with phenothiazine antiemetics treat most cases of NVP successfully.

The etiology of NVP remains uncertain, although numerous hypotheses have been proposed. Backed by the study of Fairweather et al in 1968, hysteria and other psychological factors were for many years believed to be largely responsible for NVP.[4,5] Subsequent studies, however, have not corroborated these findings. The association between NVP and esophageal and gastric dysmotility was first reported over 20 years ago. Recent studies have confirmed the association between gastric dysrhythmias and NVP and treatment of women in their first-trimester of pregnancy with high-protein meals decreased their nausea and gastric dysrhythmias.[6,7] The mechanisms of both the gastric dysrhythmias and the beneficial effects of protein meals are unknown. The proposal of hormones such as human chorionic gonadotropin (hCG), estrogen, and progesterone as etiologic agents in NVP is underscored by their rapid degree of elevation during the early phases of pregnancy. Unfortunately, multiple studies measuring hormonal levels of hCG, estrogen and progesterone in patients with NVP have yielded inconsistent results.[8] *Helicobacter pylori* (*H. pylori*) has received much recent press as a potential cause of NVP based on the findings of a study conducted in Taiwan which reported correlation between seropositivity for *H. pylori* infection and the presence of NVP.[9] The role of *H. pylori* in NVP is not well supported by the data, however, as seropositivity for infection did not correspond with the gastrointestinal symptoms of NVP.

Hyperemesis Gravidarum

HG is most severe form of NVP and affects a small but significant portion of pregnancies. HG, which occurs in approximately 1 out of every 200 pregnancies, is commonly defined as persistent first trimester vomiting that leads to weight loss in excess of 5% prepregnancy body weight as well as to ketonuria unrelated to other causes. Patients with HG are typically ill-appearing and show clinical signs of dehydration including poor skin turgor and dry mucous membranes. Jaundice is uncommon, but indicates liver involvement when present. Common laboratory derangements seen in HG include hypokalemia, hyponatremia, metabolic alkalosis, and polycythemia. Abnormal liver function tests, as discussed below, are common in patients with HG.

While HG is felt to share similar risk factors and pathogenesis with NVP as a whole, multiple gestations, molar pregnancies and fetal anomalies such as hydrops fetalis and trisomy 21 appear to impart particular risk for the development of HG.[10,11] In contrast to most patients with NVP, hospitalization for intravenous fluids, intravenous antiemetics, and bowel rest is often required in HG. Although the effects of mild cases of NVP on mother and fetus is debated, the severe vomiting and malnourishment seen in HG appears to confer poor pregnancy outcomes such as preterm labor and hemorrhage, fetal malformations, and low birth weight.[12]

Liver Disease in Hyperemesis Gravidarum

Maternal liver disease occurs in nearly 50% of patients who require hospitalization for HG.[13] In one series, HG accounted for 32% of liver function test abnormalities among pregnant women of all gestational ages and was responsible for 94% of liver function test abnormalities seen in pregnant women during their fist trimester of gestation.[14] Unfortunately, despite the high incidence of liver disease in HG, clinical methods to predict which patients with HG are at risk for liver involvement are lacking.

Spectrum of Maternal Liver Disease in Hyperemesis Gravidarum

The spectrum of liver disease in HG is broad. With the exception of rare cases of jaundice, the clinical presentation of HG with and without liver involvement is nearly identical. Mild aminotransferase elevation (up to 200 U/l) is the most common liver laboratory abnormality seen in HG, although increased alkaline phosphatase up to twice normal values and mild hyperbilirubinemia (mixed direct and indirect fractions) up to 4 mg/dl may also occur.[15] There are rare case reports of aminotransferases elevation greater than 1600 U/l, but fulminant liver failure in HG has not been reported.[16,17] The severity of nausea and vomiting in patients with HG and liver disease, as a general rule, correlates with the degree of liver enzyme elevation.[18] Imaging studies of the liver in patients with HG are usually unremarkable, although abdominal ultrasounds are always advised in patients with suspected HG in order to rule out gestational trophoblastic disease and multiple gestations. Liver biopsies are rarely necessary in HG and are performed in very few patients. When liver biopsies are done, they may show the histopathologic changes of necrosis, steatosis and bile plugs.[15,19] The treatment for HG with liver disease is the same as HG without liver disease and usually entails hospitalization for intravenous fluids, intravenous antiemetics, bowel rest and, rarely, parenteral nutrition. Unlike the other pregnancy-related liver diseases discussed elsewhere in the text, HG with liver involvement is almost never fatal. Liver function abnormalities usually return to normal levels within a few days of volume expansion and the cessation of vomiting. No long-term sequelae of liver dysfunction have been described in patients with pregnancies complicated by HG and liver disease.

Differential Diagnosis of Liver Disease in Hyperemesis Gravidarum

HG with liver disease is a diagnosis of exclusion, and the differential diagnosis of nausea and vomiting in pregnancy with abnormal liver function tests is extensive. Before the diagnosis can be made, disease states unrelated to pregnancy such as gastrointestinal disorders (e.g., viral hepatitis and cholecystitis), metabolic disease (e.g., diabetes and hyperthyroidism), and medication effects, must be ruled out. Other pregnancy-related conditions with nausea, vomiting and liver dysfunction such as the acute fatty liver of pregnancy (AFLP) and the hemolysis elevated liver enzymes and low platelets (HELLP) syndrome typically present in the third trimester of gestation and thus are readily distinguished from HG in most cases.

Pathogenesis

The Starvation Injury Hypothesis

The pathogenesis of liver disease in HG is not well understood. Initial reports by Adams et al in 1968 proposed that starvation injury might be responsible.[20] For nearly twenty years, subsequent studies indirectly supported this hypothesis. First, elevated serum aminotransferases and bilirubin were noted in patients with kwashiorkor. Second, diminished bile excretion and bile flow was observed in rat models during fasting states. Third, a temporary elevation in liver enzymes was noted in a cohort of 46 obese patients in the midst of hypocaloric "starvation" diets resulting in rapid weight loss.[21,22,23] Morali et al challenged the hypothesis of starvation injury in a retrospective review of the medical records of 80 patients hospitalized for HG between 1976 and 1986. Patients, all of whom had known urinary ketones (used as a starvation surrogate) and liver function tests, were divided into groups with either normal liver enzymes or abnormal liver enzymes. In these two groups of patients, Morali tested the statistical significance between ketonuria and elevated liver function tests and also gauged the relationship between quantitative ketonuria and the degree of liver enzyme elevation. Although ketonuria and elevated liver function tests were significantly associated, there was no correlation between the degree of ketonuria and the magnitude of liver enzyme elevation. This suggests that other factors, in addition to starvation injury, contribute to liver enzyme elevation in patients with HG.[24] Placental release of inflammatory cytokines and disorders of fatty acid oxidation may represent some of these factors.

The Placental Cytokine Hypothesis

Cytokines are soluble, low molecular weight proteins secreted by activated macrophages and lymphocytes. Cytokines have multiple biologic functions, including the regulation of immunity and inflammation. In pregnancy, the placenta is inundated with macrophages and lymphocytes and is an important source of many inflammatory cytokines including interleukin-1 (IL-1), interleukin-6 (IL-6), and tumor necrosis factor-alpha (TNF-α).[25-29] Even though cytokines are thought to be an integral part of the placental-maternal unit during normal gestation, over-expression of cytokine-producing cells has long been proposed as a potential cause of numerous pregnancy-associated diseases including preeclampsia and HG. While TNF-α-induced hepatotoxicity in viral hepatitis and alcoholic liver disease is relatively well-characterized, the hepatotoxic role inflammatory cytokines play in other conditions such as pregnancy remains unclear.[30] In order to study the role of cytokines in patients with HG, Kaplan et al measured the levels of TNF-α, IL-2, and IL-6 in 30 healthy nonpregnant women, 30 healthy pregnant women, and 30 pregnant women meeting clinical criteria for HG.[31] While there was no statistical difference between IL-2 and IL-6 levels among women with HG and the other two groups, TNF-α was found to be significantly elevated in HG patients compared with both the healthy pregnant and the healthy nonpregnant groups. Kaplan's data suggest that TNF-α may play a role in HG, although it is not clear whether cytokines are causative agents of liver disease associated with HG and liver disease or simply just by-products of the disease. Recent mouse models, however, have shown that elevated levels of TNF-α may directly lead to T-cell apoptosis within the liver which in turn could cause liver damage.[32] Moreover, liver biopsies of mouse models genetically engineered to express high levels of TNF-α showed typical features of active hepatitis including lymphocytic infiltration and necrosis.[33] While these findings are intriguing and may provide future avenues of investigation, the precise role of cytokines in HG and other pregnancy-induced liver disorders is as yet undetermined.

The Impaired Mitochondrial Fatty Acid Oxidation Defect Hypothesis

As highlighted in Chapter 7 by Choudhary et al, fatty acid oxidation (FAO) is the principal means by which humans generate energy for survival. FAO usually occurs in the mitochondria of cells but can also take place in peroxisomes. Mitochondrial FAO involves transport of free fatty acids (FFA) across selectively permeable mitochondrial membranes and requires multiple successive steps catalyzed by distinct enzymes.[34] The end products of FAO are the reduced forms of flavin adenine dinucleotide ($FADH_2$) and nicotinamide adenine dinucleotide (NADH). The subsequent electron transfer from $FADH_2$ and NADH ultimately to adenosine diphosphate (ADP) in oxidative phosphorylation leads to the production of energy-rich adenosine triphosphate (ATP).[34]

Pediatric defects in FAO are recessively inherited and have recently emerged as an important group of inborn errors of metabolism with clinical significance.[35] Affected patients with FAO defects usually present in the first year of life with a Reye's-like syndrome complicated by hypoglycemia and encephalopathy. Coma and death quickly ensue unless the disorder is rapidly recognized and treated. Conversely, patients heterozygous for mutations in the FAO cascade are usually phenotypically normal except during periods of oxidative stress such as starvation and pregnancy.

As discussed extensively in Chapter 6 by Bearelly et al and Chapter 7 by Choudhary et al, fetal FAO defects are now recognized to cause AFLP and the HELLP syndrome in the obligate heterozygous mothers during the third trimester.[36] The most commonly reported enzymatic mutation in the FAO pathway leading to pregnancy-induced liver disease is that of long chain 3-hydroxyacyl-coenzyme A dehydrogenase (LCHAD).[34] A retrospective review of the obstetric records of 29 Finnish infants diagnosed with LCHAD deficiency by Tyni et al showed that over 10% of those pregnancies were complicated by HG in the mother.[37] Fetal mutations in other enzymes of the FAO pathway leading to HG have also surfaced.[34,38] Innes et al reported a case of HG linked with fetal deficiency of hepatic carnitine palmitoyltransferase I (HCPT I), the enzyme responsible for actively transporting long chain FFA from the cytoplasm of cells across the outer mitochondrial membrane.[39]

Recent research underscores the importance of mitochondrial FAO in placental metabolism and subsequent fetal development. While maternal glycolysis was for years felt to represent the principal fetal metabolic source, Strauss and colleagues showed that the activity of the enzymes of FAO is high in placental tissues, especially during the first-trimester.[40] Furthermore, FAO-deficient mouse models have been shown to suffer from intrauterine fetal growth retardation, hypoglycemia and early death.[41]

In light of all available data, we speculate that women heterozygous for FAO defects develop HG associated with liver disease while carrying fetuses with FAO defects due to the accumulation of fatty acids in placenta and subsequent generation of reactive oxygen species. Alternatively, it is possible that starvation leading to peripheral lipolysis and increased load of fatty acids in maternal-fetal circulation, combined with reduced capacity of the mitochondria to oxidize fatty acids in mothers heterozygous for FAO defects, can also caused HG and liver injury while carrying nonaffected fetuses. This hypothesis is illustrated in Figure 2.

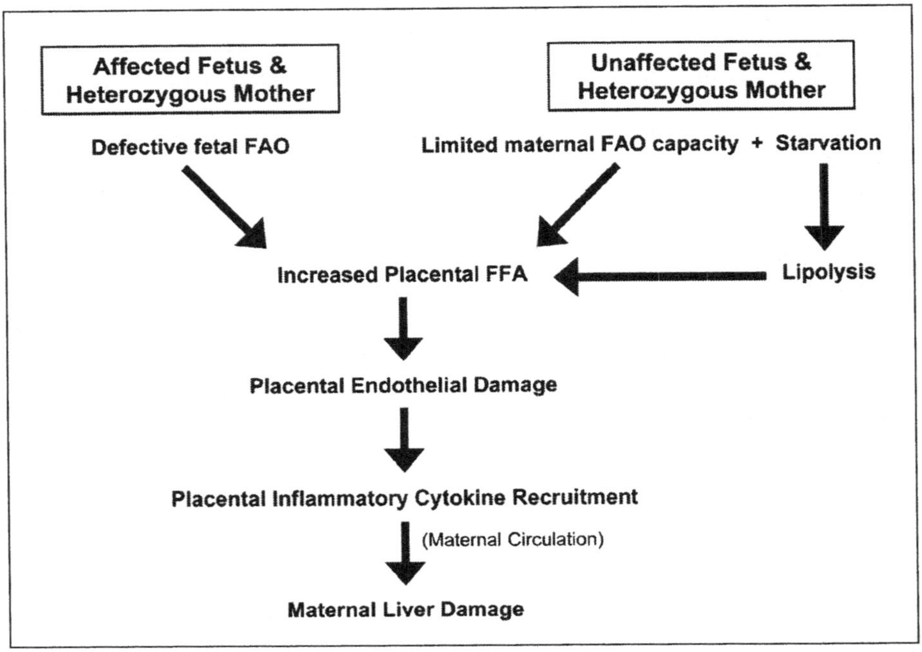

Figure 2. A unifying hypothesis for development of liver disease in HG secondary to fetal and/or maternal defect in fatty acid oxidation. Fetuses with defective FAO may lead to accumulation of fatty acid and their metabolites in the placenta with subsequent placental endothelial damage and cytokine recruitment causing maternal liver disease. Alternatively, it is possible that maternal heterozygosity for a FAO defect alone, without carrying an affected fetus, also causes placental accumulation of fatty acids and endothelial damage. This can be the result of a combination of starvation injury casing peripheral lipolysis and limited capacity of maternal mitochondria to oxidize fatty acids.

Liver Disease Associated with Hyperemesis Gravidarum and Fetal Fatty Acid Oxidation Defects

Since mitochondrial FAO is highly active in predominantly fetal-derived placental tissues,[40] FFA intermediates might accumulate in the placenta of pregnancies complicated by fetal FAO defects. Studies in nonobstetric populations have shown that FFA intermediates can lead to free-radical induced damage to cellular membranes with activation of the cytochrome P-450 and lipid peroxidation.[42,43] Fetal-derived FFA intermediates could then lead directly to placental endothelial damage followed by the recruitment of inflammatory cytokines ultimately circulated to the maternal liver. Damage to the maternal liver could follow, since inflammatory cytokines are potentially hepatotoxic.[32,33]

Liver Disease Associated with Hyperemesis Gravidarum and Nonaffected Fetuses

The starvation commonly seen in patients with HG can lead to enhanced peripheral lipolysis and an increased influx of FFA into mitochondria. Heterozygous mothers, while usually phenotypically normal, have limited capacity for mitochondrial FAO which subsequently leads to increased levels of plasma FFA intermediates and causes enhanced extramitochondrial FAO leading to the generation of reactive oxygen species. The maternal-derived FFA intermediates which accumulate in the circulation of the obligate heterozygous mothers could induce free-radical damage of placental endothelium with the subsequent release of inflammatory cytokines, resulting in maternal liver damage.

Conclusion

NVP is a common affliction during the first trimester of pregnancy, of which HG represents the most severe form. Liver involvement occurs in nearly one-half of all cases of HG and typically consists of mild elevation in serum transaminases. Liver test abnormalities resolve quickly in most cases after cessation of vomiting and volume repletion. No long-term sequelae of liver dysfunction have been described in mothers suffering with HG and liver test abnormalities. The pathogenesis of liver disease in HG is likely multifactorial. Starvation injury, placental inflammatory cytokines, and impaired FAO in either the fetus and/or mother are all likely to play a part. More research is required to help clarify the roles of the above factors, as well as the roles of as yet undetermined factors, in the etiology of the pregnancy-induced liver disease in HG.

References

1. Weigel MM, Weigel RM. Nausea and vomiting of early pregnancy and pregnancy outcome. An epidemiological study. Br J Obstet Gynaecol 1989; 96:1304-1311.
2. Broussard C, Richter J. Nausea and vomiting of pregnancy. Gastroenterol Clin North Am 1998; 27:123-151.
3. Klebanoff MA, Koslowe PA, Kaslow R et al. Epidemiology of vomiting in early pregnancy. Obstet Gynecol 1985; 66:612-616.
4. Fairweather DV. Nausea and vomiting in pregnancy. Am J Obstet Gynecol 1968; 102:135-175.
5. Quinlan JD, Hill DA. Nausea and vomiting of pregnancy. Am Fam Physician 2003; 68:121-128.
6. Koch KL, Stern RM, Vasey M et al. Gastric dysrhythmias and nausea of pregnancy. Dig Dis Sci 1990; 35:961-968.
7. Jednak MA, Shadigian EM, Kim MS et al. Protein meals reduced nausea and gastric slow wave dysrhythmic activity in first trimester pregnancy. Am J Physiol 1999; 277:G855-861.
8. Goodwin TM. Hyperemesis gravidarum. Clin Obstet 1998; 41:597.
9. Wu CY, Tseng JJ, Chou MM et al. Correlation between Helicobacter pylori infection and gastrointestinal symptoms in pregnancy. Advan Ther 2000; 17:152-158.
10. Philip B. Hyperemesis gravidarum: Literature review. WMJ 2003; 102:46-51.
11. Koch KL, Frissora CL. Nausea and vomiting during pregnancy. Gastroenterol Clin Am 2003; 32:201-234.
12. Tsang I, Katz V, Wells S. Maternal and fetal outcomes in hyperemesis gravidarum. Int J Gynecol Obstet 1996; 55:231-235.
13. Abell T, Riely C. Hyperemesis gravidarum. Gastroenterol Clin North Am 1992; 21:835.
14. Wong HY, Tan JYL, Lim CC. Abnormal liver function tests in the symptomatic pregnant patient: The local experience in Singapore. Ann Acad Med 2004; 33:204-208.
15. Larry D, Rueff B, Feldmann G et al. Recurrent jaundice caused by recurrent hyperemesis gravidarum. Gut 1984; 25:1414.
16. Conchillo JM, Koek GH. Hyperemesis gravidarum and severe liver enzyme elevation. J Hepatol 2002; 37:162-166.
17. Orazi G, Dufour PH, Puech F. Jaundice induced by hyperemesis gravidarum. Int J Gynecol Obstet 1998; 61:181-183.
18. Bacq Y, Funai EF, Riely CA. Hyperemesis gravidarum. Up To Date. www.utdol.com.
19. Knox TA, Olans LB. Liver disease in pregnancy. N Engl J Med 1996; 335:569-576.
20. Adams RH, Gordon J, Combes B. Hyperemesis gravidarum. I Evidence of hepatic dysfunction. Obstet Gynecol 1968; 31:659.
21. Webber B, Freiman I. The liver in kwashiorkor. Arch Pathol 1974; 98:400-408.
22. Mahn J, Duraldestin P, Dhumeaux D et al. Biliary transport of cholephilic dyes: Evidence for two different pathways. Am J Physiol 1977; 232:E445-450.
23. Friis R, Vaziri D, Akbarpour F et al. Effect of rapid weight loss supplemented fasting on liver tests. J Clin Gastroenterol 1987; 9:204-207.

24. Morali GA, Braverman DZ. Abnormal liver enzymes and ketonuria in hyperemesis gravidarum. J Clin Gastroenterol 1990; 12:303-305.
25. Romero R, Mazor M. Infection and preterm labor. Clin Obstet Gynecol 1988; 31:553-584.
26. Kameda T, Matsuzaki N, Sawai K et al. Production of interleukin-6 by normal human trophoblast. Placenta 1990; 11:205-213.
27. Li Y, Matsuzaki N, Masuhiro K et al. Trophoblast-derived tumor necrosis factor-alpha induces release of human chorionic gonadotropin using interleukin-6 and IL-6-receptor-dependent system in the healthy human trophoblasts. J Clin Endocrinol Metab 1992; 74:184-191.
28. Casey ML, Cox SM, Beutler B et al. Cachectin/tumor necrosis factor-alpha formation in human deciduas: Potential role of cytokines in infection-induced preterm labor. J Clin Invest 1989; 83:430-436.
29. Romero R, Manogue KR, Mitchell MD et al. Cachectin-tumor necrosis factor in the amniotic fluid of women with intraamniotic infection and preterm labor. Am J Obstet Gynecol 1989; 161:336-341.
30. Neuman MG. Cytokines—central factors in alcoholic liver disease. Alcohol Res Health 2003; 27:307-316.
31. Kaplan PB, Gucer F, Sayin NC et al. Maternal serum cytokine levels in women with hyperemesis gravidarum in the first trimester of pregnancy. Fertil Steril 2003; 79:498-502.
32. Murray DA, Crispe IN. TNF-alpha controls intrahepatic T-cell apoptosis and peripheral T-cell numbers. J Immunol 2004; 173:2402-2409.
33. Mohammed FF, Smookler DS, Taylor SE et al. Abnormal TNF activity in Timp3-/- mice leads to chronic hepatic inflammation and failure of liver regeneration. Nat Genet 2004; 36:969-977.
34. Rakheja D, Bennett MJ, Rogers BB. Long-Chain L-3-Hydroxyacyl-Coenzyme a dehydrogenase deficiency: A molecular and biochemical review. Lab Invest 2002; 82:815-824.
35. Ibdah JA, Yang Z, Bennett MJ. Liver disease in pregnancy and fetal fatty acid oxidation defects. Mol Genet Metab 2000; 71:182-9.
36. Schoeman MN, Batey RG, Wilcken B. Recurrent acute fatty liver of pregnancy associated with a fatty-acid oxidation defect in the offspring. Gastroenterology 1991; 100:544-548.
37. Tyni T, Ekholm E, Pihko H. Pregnancy complications are frequent in long-chain 3-hydroxyacyl-coenzyme a dehydrogenase deficiency. Am J Obstet Gynecol 1998; 178:603-608.
38. Ibdah JA, Bennett MJ, Rinaldo P et al. A fetal fatty-acid oxidation disorder as a cause of liver disease in pregnant women. N Engl J Med 1999; 340:1723-1731.
39. Innes AM, Seargeant LE, Balachandra K et al. Hepatic carnitine palmitoyltransferase I deficiency presenting as a maternal illness in pregnancy. Pediatr Res 2000; 47:43-45.
40. Shekhawat P, Bennett MJ, Sadovsky Y et al. Human placenta metabolizes fatty acids: Implications for fetal fatty acid oxidation disorders and maternal liver diseases. Am J Physiol Endocrinol Metab 2003; 284:E1098-E1105.
41. Ibdah JA, Paul H, Zhao Y et al. Lack of mitochondrial trifunctional protein in mice causes neonatal hypoglycemia and sudden death. J Clin Invest 2001; 107:1403-1409.
42. Mak IT, Kramer JH, Weglicki WB. Potentiation of free radical-induced lipid peroxidative injury to sarcolemmal membranes by lipid amphiphiles. J Biol Chem 1986; 261:1153-1157.
43. Singh AK, Yoshida Y, Garvin AJ et al. Effect of fatty acids and their derivatives on mitochondrial structures. J Exp Pathol 1989; 4:9-15.

Chapter 5

Intrahepatic Cholestasis of Pregnancy

Paul A. Dawson* and Girish Mishra

Abstract

Intrahepatic cholestasis of pregnancy (ICP), also known as obstetric cholestasis, is one of the most common causes of liver disease in pregnancy. ICP is a reversible form of cholestasis that usually develops in the third trimester of pregnancy and is associated with pruritus, elevated serum levels of bile acids, abnormal liver function tests, and mild changes in liver histology. These symptoms persist until delivery, after which the condition rapidly resolves. ICP is largely a clinical diagnosis and requires a suggestive history and exclusion of other causes of cholestasis. Although ICP is considered a benign disease with a good prognosis for the mother, ICP is associated with poor fetal outcomes and perinatal mortality as high as 15% in untreated cases and 3% in actively managed cases. The etiology of ICP is unclear, but undoubtedly involves genetic and hormonal factors. Genetic mutation and association studies have implicated genes for the liver canalicular phospholipid transporter MDR3 (*ABCB4*), the canalicular bile acid transporter BSEP (*ABCB11*), the canalicular organic anion transporter MRP2 (*ABCC2*), and the bile acid-activated nuclear receptor FXR. However, most studies suggest that ICP is a genetically heterogeneous disease and these individual loci contribute to only a fraction of the ICP cases. Current evidence suggests that ICP involves abnormal metabolism of progesterone and bile acids that leads to elevated maternal serum bile acids and impaired transport of bile acids across the placenta from the fetal to the maternal circulation. However, it is not known how this increased fetal bile acid burden leads to the ICP-associated complications such as perinatal mortality, preterm delivery, abnormal intrapartum fetal heart rate rhythm, and meconium passage. Ursodeoxycholic acid (UDCA) is currently the most efficacious medical intervention. UDCA therapy relieves maternal symptoms, reduces serum liver enzymes and, more importantly, may also protect the fetus from bile acid toxicity and improve fetal survival. Due to the increased risk of stillbirth in ICP, most experienced centers advocate delivery by the 37th to 38th week of gestation.

Introduction

Intrahepatic cholestasis of pregnancy (ICP), also known as obstetric cholestasis, is one of the most common causes of liver disease in pregnancy.[1-4] ICP is a reversible form of cholestasis that usually develops in the third trimester of pregnancy and is associated with itching, elevated serum levels of bile acids, abnormal liver function tests, and mild changes in liver histology. These symptoms persist until delivery, after which the condition rapidly resolves. Although

*Corresponding Author: Paul A. Dawson—Division of Gastroenterology, Department of Internal Medicine, Wake Forest University School of Medicine, Winston-Salem, North Carolina, USA. Email: pdawson@wakehealth.edu

Maternal Liver Disease, edited by Jamal A. Ibdah. ©2012 Taylor & Francis Group, LLC

ICP is considered a benign disease with a good prognosis for the mother, ICP is associated with poor fetal outcomes. Clinical studies have shown that the major consequences of ICP are premature births (19 to 60%), fetal distress (22 to 41%), and intrauterine fetal death (0.75 to 1.6%) in the affected pregnancies.[1-3]

The incidence of ICP has been reported to be between ~10 and 1500 cases per 10,000 pregnancies, reflecting the striking geographical and ethnic differences.[2] ICP is most common in Scandinavia and South America, with particularly high rates reported in Chile and Bolivia.[5,6] In Chile, the rate of ICP was originally reported as high as 16% in the general population and 28% in women of Araucanian Indian descent. However, this study did not fully differentiate between ICP and the more benign condition of pruritus gravidarum,[5] and more recent reports from Chile suggest a prevalence for ICP closer to 2 to 4%.[7] ICP is less common in Europe, the United States, Canada, and Australia, with a reported prevalence between 0.1 to 1.5%.[2] However, it must be noted that the diagnostic criteria for ICP has varied over time and this is reflected in the different studies. While pruritus remains the common diagnostic factor, pruritus in itself is common in pregnancy where the reported incidence ranges from 3% to as high as 14% of women in early pregnancy.[8] Thus, it is important to distinguish pruritus associated with ICP and poor fetal outcomes from more benign pruritus-associated conditions. The terms 'obstetric cholestasis', 'recurrent jaundice of pregnancy', 'pruritus gravidarum', 'icterus gravidarum', and 'idiopathic jaundice of pregnancy' all refer to a spectrum of conditions characterized by pruritus with or without jaundice due to cholestasis, and it has been suggested that they represent different stages of severity of the same clinical entity.[1]

Clinical Features

ICP typically occurs in the third trimester of gestation (week 20 to 35) but may begin earlier.[9,10] Onset during the first and second trimesters has been reported in 10% and 25% of ICP cases, respectively.[1] Pruritus is the most common presenting symptom and is more severe at night and on the palms and soles. Although a benign condition, the pruritus can be debilitating. Pruritus usually starts in the third trimester, persists until the end of gestation, and disappears a few days after delivery. Jaundice is less common and develops in only about 10 to 15% of patients.[11] The jaundice is usually mild and develops after the onset of pruritus.[12] Although less common, mild steatorrhea, malabsorption of fat-soluble vitamins, and weight loss have also been reported in ICP.[13] In those rare cases of severe steatorrhea, it has been suggested that patients be supplemented with vitamin K before delivery to mitigate potential risks of intra- and postpartum hemorrhage.[9,13] Abdominal pain is not a feature and liver failure does not occur in ICP.

Women with ICP are at increased risk for gallstone formation and cholecystitis (rate ratio of 3.7),[14] and conversely, a history of gallstone disease correlates with a higher incidence of ICP.[15,16] For example, in a large prospective study conducted by Glantz et al, the prevalence of gallstone disease (defined as prior cholecystectomy or ultrasound-verified gallstones) was 0.5% in pregnant women without ICP versus 2.3% in women with mild ICP and 7.4% in women with severe ICP.[16] In addition to an increased risk of gallstones during pregnancy, there is an increased incidence of cholelithiasis after delivery for mothers who developed ICP. Using ultrasonography, gallstones were more likely to be found in primigravida women with ICP versus those without ICP (12% versus 7.7%).[17]

Symptomatic relief and improvement in the laboratory tests begins with delivery and are usually prompt. By 4 to 6 weeks after delivery, the biochemical aberrations typically return to normal values.[1] Cholestasis is rarely prolonged in ICP patients,[14,18] and this occurrence should suggest an alternative diagnosis. Other causes of pruritus and jaundice require exclusion, especially gallstones, primary biliary cirrhosis (PBC), primary sclerosing cholangitis (PSC), viral hepatitis, autoimmune chronic active hepatitis, and drug/dietary supplement-associated hepatotoxicity. ICP is common in subsequent pregnancies, re-occurring with varying disease severity in most women.[19]

Fetal Outcomes

There is a well-known but poorly understood association between ICP and poor perinatal outcome. The fetus fares less well than the mother and there are many reports of an increased frequency of fetal distress, unexplained stillbirth, and need for premature delivery and neonatal care.[1,2] Although the fetal risks may be minimized by heightened surveillance and close monitoring of an affected mother in experienced centers,[2] historically ICP is associated with a variety of adverse fetal outcomes including meconium staining of amniotic fluid[20,21], asphyxia events,[16] fetal brachycardia and tachycardia,[20,22,23] and fetal loss that tends to cluster around weeks 37 to 39.[2] The risk of adverse fetal events appears to be related to elevated maternal serum bile acids levels. Fetal complications such as, asphyxia events and meconium staining of amniotic fluid were not observed until the maternal fasting serum bile acids levels were $\geq$40 μmol/L.[16] Similarly, the risk of premature delivery is significantly increased in ICP with reported rates of 30 to 40%.[11,19,20] Premature delivery in ICP is predicted by early onset of pruritus and higher maternal fasting serum bile acids.[24] The mechanisms are unclear, but it is hypothesized that the high incidence of spontaneous preterm labor is partially due to the significant increase in uterine contractile activity noted during the course of the disease.[25] As described below, bile acids have been implicated in the pathogenesis of the fetal complications.[2,26]

Diagnosis

ICP occurs in women of all maternal ages and in both primiparous and multiparous women. The disorder may recur with subsequent gestations, and 60 to 70% of patients affected in their initial pregnancy have recurrence. Interestingly, the recurrence may not be as severe as the first episode.[19] The diagnosis of ICP is made primarily on a clinical basis and exclusion of other causes by the history, serum markers, and upper abdominal ultrasound.

Liver Function Tests

In general, pregnant women who experience pruritus should have liver function tests. In ICP, laboratory testing confirms the presence of cholestasis, with elevations in fasting serum bile acids of greater than 10 μmol/L in most patients. The elevation of serum bile acids levels precedes abnormalities of other liver tests,[27,28] although pruritus might precede any abnormalities in liver tests including serum bile acids.[29,30] Among standard liver tests, alanine aminotransferase (ALT) is a more sensitive marker for ICP.[9,28,31] Since serum ALT and AST levels are normal until delivery in healthy pregnancies, any rise in their levels should lead to a heightened suspicion for ICP. Between 20 and 60% of women with pruritus and elevated serum bile acids have a 2- to 10-fold rise in serum transaminases.[27] A positive correlation between the levels of transaminases and bile acids has also been reported in some[32] but not all studies.[31,33] The serum levels of alkaline phosphatase (AP) are elevated in normal pregnancies due to higher production of placental and bone isoenzymes. As such, the elevated levels of AP found in ICP patients are of poor prognostic value. In contrast to AP, serum gamma glutamyl transpeptidase (GGT) levels are normally lower in late pregnancy.[34] For ICP, serum GGT levels have been reported to be elevated in only about a third of the cases.[31,35] Interestingly, this enzymatic picture of a normal GGT and cholestasis is similar to the presentation in pediatric patients with certain forms of progressive familial intrahepatic cholestasis such as Byler's disease.[9] As mentioned above, significant hyperbilirubinemia is only detected in about 10% of cases. This is consistent with reports that total and free bilirubin concentrations are reduced in pregnancy[34] and only about 0.07% of all pregnant women develop visible jaundice.[36] The increase in plasma levels of low density lipoprotein (LDL)-cholesterol and triglyceride, which is normally observed during pregnancy,[36] is more pronounced in cholestatic patients including women with ICP.[37] Cholestasis is associated with the appearance of an abnormal lipoprotein, termed lipoprotein X or LpX, in plasma.[38,39]

Serum Bile Acids

Elevation of fasting serum primary bile acids such as cholic or chenodeoxycholic acid is considered the most sensitive laboratory indicator for diagnosis of ICP. Although maternal fasting serum bile acid levels (typically 5-6 μmol/L) are unchanged or only slightly higher in normal pregnancies,[34,40] they can be elevated as much as 10 to 100-fold in ICP. In a landmark prospective study, 45,485 pregnancies in Sweden were screened for ICP to determine the incidence and fetal complication rates.[16] As mentioned above, increased risk of fetal complications was detected in patients with serum bile acid concentrations ≥ 40 μmol/L, but no increased fetal risk was observed in ICP patients with bile acids levels ≤ 40 μmol/L.

Although elevated serum bile acid levels and abnormal liver tests are useful markers for ICP in late pregnancy, these markers have not proven useful for distinguishing the benign condition of pruritus gravidarum from ICP in early gestation. For example, results from a prospective longitudinal study showed that serum bile acids, ALT, and AST were not significantly elevated in ICP patients until gestational week 24 or later. However, LDL and serum glutathione S-transferase (GST)-alpha levels were already significantly higher at 16 weeks in women with ICP as compared to women with pruritus gravidarum or normal pregnancies.[32,37] The longitudinal profiles of maternal serum bile acids have recently been documented between gestational week 16 and 4 weeks postpartum in normal pregnancy and in pregnancies complicated by ICP and pruritus gravidarum.[33] Following week 24 of pregnancy, serum bile acid levels were elevated and profiles altered in ICP patients versus normal pregnancy and pruritus gravidarum.

Histology

Liver biopsies are generally not warranted in patients with ICP except in atypical cases such as when symptoms start before gestational week 20, jaundice precedes pruritus, or itching persists following delivery. The histological presentation of ICP typically includes mild focal irregular intrahepatic cholestasis ("bland cholestasis"; i.e., without any parenchymal inflammation) with bile plugs in the canaliculi and absent or only mild portal tract inflammation. There may also be widening of the bile canaliculi, as demonstrated using electron microscopy.[41,42] However, biopsy is typically of little diagnostic value since this "bland" cholestasis is not unique to ICP and is found in a variety of cholestatic disorders including progressive familial intrahepatic cholestasis Type 1 (PFIC1), benign recurrent cholestasis, cystic fibrosis, Dubin-Johnson syndrome in the neonate, and Crigler-Najjar syndrome.[43,44]

Differential Diagnosis

The main differential diagnoses of ICP without jaundice include specific dermatoses of pregnancy, allergic reactions, and pruritus associated with striae gravidarum.[45] Specific dermatoses of pregnancy, such as prurigo of pregnancy, can easily be ruled out by the absence of primary skin lesions in ICP. In addition, ICP with abnormal liver tests and/or jaundice should be distinguished from acute fatty liver of pregnancy, preeclampsia complicated by liver enzymes or HELLP syndrome (characterized by hemolysis, elevated liver enzymes, and low platelets), hyperemesis gravidarum, viral hepatitis, autoimmune hepatitis, parasitic infections, skin allergies, and other cholestatic states such as PBC, PSC, benign recurrent cholestasis, and obstructive biliary disease.[1] If cholestasis and elevated transaminases persist for more than 4 weeks after delivery, chronic liver diseases such as PBC or PSC have to be ruled out using appropriate serologies and imaging modalities.[46]

Genetics

Mode of Inheritance

A genetic predisposition for ICP is supported by the finding of a heterogeneous incidence in women of different ethnic origin,[1] familial clustering as documented in several pedigree studies,[47,48] and higher incidences in mother and sisters of patients with ICP.[49,50] However, subsequent studies to elucidate the genetic basis of ICP have met with only mixed success. Despite some initial suggestive results,[48] most studies have failed to demonstrate any relationship between HLA haplotypes and ICP.[51-53] Although analysis of various pedigrees generally favors an autosomal dominant inheritance,[47,48,52,54] the lack of a male phenotype, incomplete penetrance, and the requirement of pregnancy for diagnosis has complicated attempts to establish the mode of inheritance for ICP. For example in a study of 11 Finnish ICP families, 3 pedigrees showed a typical dominant inheritance while analysis of the remaining 8 pedigrees was inconclusive.[54]

Candidate Genes

In addition to on-going association and linkage studies, numerous groups have begun to evaluate candidate genes for ICP. This approach was spurred on in part by successes in identifying genes responsible for Progressive Familial Intrahepatic Cholestasis (PFIC) Types 1, 2, and 3.[55,56] PFIC Type 1 (formerly called Byler's disease) is a chronic familial intrahepatic cholestasis in patients with normal serum GGT levels. An extensive search for the disease loci identified a P-type ATPase, designated FIC1 (gene symbol, *ATP8B1*), as the defective gene in PFIC1 as well as in benign recurrent intrahepatic cholestasis (BRIC).[57] An analysis of the spectrum of *ATP8B1* mutations in PFIC1 and BRIC patients revealed that mutation type and location in the gene generally correlates with clinical severity; more benign missense mutations are common in BRIC while nonsense, frame shifting, and deletion mutations are more common in PFIC1.[58] A second form of familial intrahepatic cholestasis, PFIC2 is associated with progressive cholestasis, lobular and portal fibrosis, giant cell transformation, normal serum GGT levels, and no bile duct proliferation. The PFIC2 gene mapped to chromosome 2q24 and was shown to be the canalicular Bile Salt Export Pump (BSEP; gene symbol *ABCB11*) that is responsible for hepatic bile acid secretion.[59] In addition to a progressive intrahepatic cholestatic phenotype, mutations in *ABCB11* have been found in patients with a milder phenotype similar to benign recurrent intrahepatic cholestasis.[60] Based on these results and consistent with the genetic classification of PFIC into subtypes, this disorder is now classified as BRIC Types 1 and 2 for the *ATP8B1* and *ABCB11* mutation-associated forms, respectively.[60] A third form of progressive familial intrahepatic cholestasis, PFIC3 is quite different from the other PFIC subtypes. Serum GGT levels are markedly elevated in these patients, and liver histology shows extensive bile duct proliferation and portal and periportal fibrosis. The gene defect responsible for PFIC3 is MDR3 (gene symbol *ABCB4*), the canalicular phosphatidylcholine (PC) transporter.[61] In PFIC3, hepatic bile acid secretion is unimpaired, but PC transport is greatly diminished. In bile, PC normally forms mixed micelles with bile acids and acts to buffer their cytotoxic detergent properties.[62] In the absence of biliary phospholipid, the bile acid monomers are highly toxic and cause a cholestatic type of liver damage.

There are intriguing similarities as well as significant differences between ICP and PFIC. Notably, a higher incidence of ICP or ICP-like symptoms has been observed in mothers of patients with PFIC or BRIC.[63,64] These observations led a number of groups to examine the PFIC genes in patients with ICP. The most promising results have been obtained for MDR3 (*ABCB4*), the gene that is defective in PFIC3.[61,65] The initial report described an inbred family in which homozygosity for a 1-base pair deletion in the MDR3 gene was associated with severe intrahepatic cholestasis necessitating a liver transplant at the age of 6 years. Heterozygosity for this mutation in 6 female relatives was associated with at least one episode of ICP, as characterized by pruritus, abnormal serum liver tests with a cholestatic pattern in the third trimester, and fetal death in 3 of the women.[65] Notably, cholestasis disappeared spontaneously and progressively after delivery.

Although that work supported an association between *ABCB4* mutations and ICP, a subsequent study of eight women with ICP, elevated serum GGT and no family history of PFIC identified a heterozygous MDR3 mutation in only one of the eight patients.[66] Some additional studies did identify potentially dysfunctional MDR3 mutations in ICP patients, but the number of affected patients with mutations tends to be small.[67,68] For example, only two out of 184 ICP patients were identified with MDR3 mutations in a study from the UK[69] and one out of 20 ICP patients in a study from Chile.[70] Similar disappointing results have been obtained for the MDR3 gene in Scandinavian ICP patients, where linkage, haplotype, and direct sequencing analysis revealed no evidence for involvement of *ABCB4*.[54,71] Overall, the results support the hypothesis that the etiology of ICP is heterogeneous and that only a small fraction of ICP can be accounted for by *ABCB4* mutations.

The PFIC1 (*ATP8B1*) and PFIC2 (*ABCB11*) genes have also been examined in ICP patients. As noted above, a higher incidence of ICP or ICP-like symptoms was observed in mothers of patients with PFIC or BRIC.[63,64] However, unlike *ABCB4*, the evidence is weaker for a direct role of FIC1 (*ATP8B1*) or BSEP (*ABCB11*) mutations in the etiology of ICP. Many studies of the BSEP gene and ICP have failed to find any association,[54,72,73] despite some early suggestive results.[74,75] A larger and more recent study of several large European ICP cohorts identified *ABCB11*-associated mutations in a small proportion of ICP cases, concluding that heterozygosity for the common *ABCB11* mutations accounts for about 1% of the ICP cases and that the V444A polymorphism in BSEP appeared to be a significant risk factor for ICP in this population.[76] The evidence for any association between the PFIC1 gene (*ATP8B1*) and ICP is similarly conflicting. A Scandinavian study that screened *ATP8B1* in 176 familial and sporadic ICP cases failed to find any evidence of *ATP8B1* mutations or a familial association between *ATP8B1* and ICP.[54,77] A study from the UK sequenced all the *ATP8B1* coding exons and identified nonsynomous heterozygous amino acid changes (D70N, R867C) in 3 of 16 ICP patients, but subsequently screening failed to identify these mutations in a parallel set of 182 women with ICP.[75] Other candidate genes that have been examined in ICP include *ABCC2* and FXR.[78-80] Mutations in the organic anion and conjugate transporter MRP2 (*ABCC2*) are known to cause Dubin-Johnson syndrome, an autosomal recessive disorder characterized by benign conjugated hyperbilirubinemia.[81] Although jaundice is exacerbated during pregnancy in Dubin-Johnson syndrome patients, fetal mortality does not appear to be increased.[82,83] To date, no specific mutations were identified in *ABCC2* in ICP patients. There are conflicting results regarding the association between single nucleotide polymorphisms in *ABCC2* and risk of ICP.[78,79] For FXR, rare mutations were identified in several cohorts of ICP patients and at least one variant was found to be significantly associated with ICP.[80] In summary, although a strong genetic predisposition is clearly evident for ICP, the identity of most of the susceptibility genes remains to be elucidated.

Molecular Pathogenesis

The etiology of ICP is still unknown, although both genetic predisposition and hormonal factors have been strongly implicated in the pathogenesis. Most studies suggest that abnormal hepatic metabolism and/or hepatic biliary secretion of sex hormones and bile acids play a critical role in ICP.[84] However, the molecular mechanisms responsible for the disorder remain to be elucidated.

Historically, ICP has been associated with the cholestatic effect of estrogen metabolites, and more recently with changes in progesterone metabolism. Several clinical observations initially implicated sex hormones in the pathogenesis of ICP.[85] First, ICP typically appears in the third trimester, a time when hormone concentrations are highest. Second, there is an increased incidence of ICP in twin pregnancies, a state with higher hormone levels.[86] Third, ICP resolves promptly after delivery, when hormone levels return to normal. Fourth, symptoms such as pruritus may re-occur in affected women when taking oral contraceptives.[87] Despite these associations, it should be noted that the total amount of estrogens and progesterone circulating in the blood or excreted in the urine of ICP patients is similar to that in normal pregnancies. Thus, the search has focused on differences in the profiles for hormone metabolites.

Several lines of evidence support a potential role for hormone metabolites in the pathogenesis of ICP. Estrogens, in particular glucuronides like estradiol-17β-glucuronide were found to be cholestatic in animal studies where they inhibited canalicular anion secretion through several different mechanisms. The canalicular multi-specific organic anion conjugate transporter, MRP2 (gene name, *ABCC2*) is responsible for the secretion of estradiol-17β-glucuronide, and this gene is down regulated in animal models of estrogen-induced cholestasis.[88] Secondly, the newly secreted estradiol-17β-glucuronide has been shown to cause a *trans*-inhibition of the canalicular bile acid export pump (BSEP).[89] In addition to estrogens, progesterone metabolites may play a particularly important role.[84] Progesterone treatment during the third trimester is associated with ICP and the profile of progesterone metabolites in plasma from patients with ICP is markedly different from that seen in normal pregnancy.[84] The question arises as to whether the changes can be explained on the basis of cholestasis alone or by specific defects in progesterone metabolism or sulfation. In this regard, it should be noted that pregnant women with cholestasis due to viral hepatitis have a normal profile of steroid sulfates in blood, suggesting that the altered pattern of sulfated progesterone metabolites is specific to ICP.[90] Although the evidence thus far remains indirect, the primary defect in ICP may be a change in the metabolism of progesterone.[91,92] Additional recent support for this idea comes from animal studies. The progesterone metabolite, 17-hydroxyprogesterone was found to accumulate in a mouse model of cholestatic liver injury, where the nuclear receptor FXR and small heterodimer partner (SHP) were both inactivated. Interestingly, administration of 17-hydroxyprogesterone alone to normal mice was sufficient to induce liver injury.[93]

ICP is associated with fetal arrhythmia and exciting new results have identified potential mechanisms by which bile acids may regulate cardiovascular function under physiological and pathophysiological conditions such as cirrhosis and ICP.[94-97] In a model using cultured neonatal rat cardiomyocytes, taurocholate was shown to act as a partial agonist for the muscarinic M2 receptor to induce desynchronization of calcium-ion fluxes, a characteristic of arrhythmias.[96] The taurocholate-induced desynchronization appears to be propagated by myofibroblasts, which transiently appear in fetal ventricular tissue during the second and third trimester of gestation. This correlates with the period when ICP-related fetal death occurs, and the absence of cardiac myofibroblasts may explain why the fetal but not maternal heart is susceptible to taurocholate-induced arrhythmia.[97] Notably, ursodeoxycholic acid (UDCA) was shown to protect against the arrhythmia-inducing effects of taurocholate, providing a potential explanation for part of the beneficial effects of UDCA in ICP.[97]

Management and Treatment Options

Maternal outcomes are generally good with pruritus and jaundice resolving soon after delivery. A variety of medications including bile acid sequestrants such as cholestyramine, antihistamines, and opioid antagonists have been used with mixed success to alleviate maternal pruritus. However, none of these agents has been shown to alter enzyme abnormalities or impact fetal outcomes. Phenobarbital, charcoal, ultraviolet light, evening primrose oil, and epomediol have all yielded disappointing results. Epomediol is a terpenoid compound that prevents and reverses cholestasis induced by ethinyl estradiol in the rat. In patients with ICP, epomediol alleviates pruritus but does not alter either liver enzymes or improve fetal prognosis.[98] The use of phenobarbital and antihistamines are discouraged since they may aggravate the respiratory difficulties of a preterm baby by causing sedation. The exciting recent finding that the albumin-bound lipid, lysophosphatidic acid (LPA), is a major mediator of cholestatic pruritus offers new potential avenues for therapeutic intervention in ICP and other cholestatic disorders.[99] LPA is generated from lysophosphatidylcholine through the action of the serum enzyme autotaxin. As circulating autotaxin levels appear to correlate with the intensity of pruritus in cholestatic patients and autotaxin inhibitors have been developed as potential anti-inflammatory and anti-cancer agents, it will be important to explore the role of LPA and potential benefit of targeting autotaxin in the treatment of cholestasis-associated pruritus.[100,101]

Cholestyramine

Cholestyramine is an exchange resin that binds bile acids and other anions in the intestine and increases their fecal excretion. Although recent guidelines continue to recommend the use of cholestyramine to treat cholestasis-associated pruritus,[102] it does not improve biochemical parameters or fetal outcomes in ICP.[103] An untoward effect of cholestyramine is the potential of increasing postpartum hemorrhage. Cholestyramine may actually worsen maternal and fetal outcome by accelerating the malabsorption of maternal vitamin K, leading to coagulopathies and, in case reports, fetal cerebral hemorrhage.[104] Similar to cholestyramine, the clinical results for the use of S-adenosyl-methionine (SAMe) in ICP are disappointing. In animal models, SAMe had been shown to decrease the negative effects of ethinyl estradiol on bile flow and increase the sulphates available for the detoxification of bile acids. However, randomized studies evaluating SAMe have yielded conflicting results with limited efficacy.[105-107]

Ursodeoxycholic Acid

Ursodeoxycholic acid (UDCA) is a naturally occurring hydrophilic bile salt and currently is first-line therapy for ICP. UDCA is approved for use in the treatment of Primary Biliary Cirrhosis and has been widely used in the management of other forms of cholestasis.[108-110] Evidence in support of the use of UDCA in ICP comes from a series of clinical trials, as well as from animal models of ICP. For example, a randomized, double-blind, placebo-controlled study of UDCA treatment in ICP demonstrated not only improvement for pruritus and liver enzymes, but also delivery significantly closer to term (37.8 ± 0.9 weeks compared with 33.8 ± 7.1 weeks).[111] Similar improvements in the symptoms and liver enzymes were seen in a placebo-controlled study of 16 ICP patients treated with UDCA or placebo.[112] Notably in that study, all the UDCA treated patients went to 38 weeks whereas the placebo-treated patients delivered at around 34 weeks, resulting in lower birth weights. In more recent trials, the efficacy of UDCA in ICP has been compared with that of SAMe,[113] cholestyramine,[114] and dexamethasone.[115] In a randomized trial of UDCA versus SAMe in ICP, women receiving UDCA had a significantly greater decreases in serum bile acids, serum ALT and AST activities, serum bilirubin, although both UDCA and SAMe were similarly effective for pruritus.[113] In a comparison of UDCA and cholestyramine, a group of 84 symptomatic patients with ICP were randomized to receive UDCA or cholestyramine for 14 days and the effects on pruritus, and pregnancy outcomes were examined.[114] In that open-label, parallel group study, UDCA was superior to cholestyramine at reducing pruritus (66.6% versus 19.0 %), reducing serum ALT and AST activities (>73% versus 21%), reducing serum bile acids levels (59% versus 19%), and time of delivery (38.7 versus 37.4 weeks). In the trial of UDCA versus dexamethasone, 130 women with ICP were randomized to UDCA, dexamethasone, or placebo and the effects on pruritus, serum liver tests, and fetal complications were examined. Intention to treat analysis showed UDCA was more effective than dexamethasone in alleviating pruritus, and reducing serum bile acids, aminotransferases, and bilirubin versus dexamethasone or placebo in women with serum bile acids levels exceeding 40 μmol/L at inclusion to the study.[115] The question of whether UDCA should be used in combination with SAMe has also been examined in several trials. Although a small placebo-controlled single blinded study of 32 ICP patients treated with

UDCA, SAMe, or UDCA plus SAMe suggested that the combination of UDCA and SAMe was superior to either agent alone,[116] an additive effect of SAMe versus UDCA was observed in a more recent comparative study.[117]

Despite extensive study, the exact mechanisms responsible for the beneficial effects of UDCA are still being elucidated. UDCA is known to modify the bile acid pool composition by replacing more cytotoxic hydrophobic bile acids and to promote hepatic bile secretion by increasing canalicular expression of transporters such as BSEP, thereby aiding in the removal of more toxic bile acids from fetal circulation.[118,119] Indeed, UDCA therapy has been shown to reduce serum taurocholate and taurodeoxycholate levels in patients with ICP.[33,120] However, as demonstrated in recent studies demonstrating UDCA protection against bile acid-induced fetal arrhythmias and formation of syncytial knots in placental tissue,[97] UDCA may also act through other mechanisms.

The US Food and Drug Administration (FDA) includes UDCA in Pregnancy Category B. Although no teratogenicity has been shown in animal studies with UDCA, there have been reports of cytotoxicity and embryonic defects in animals receiving lithocholic acid, a naturally occurring secondary bile acid.[121] However, since one of the proposed mechanisms of action for UDCA is to decrease the amount of lithocholic acid in the fetal circulation, teratogenicity concerns for UDCA may be misplaced. In addition, UDCA would not be prescribed to the mother until the third trimester when embryogenesis is complete. In fact, a 12-year experience treating ICP patients with UDCA in the third trimester found that UDCA facilitated deliveries at term, with higher birth weight than historical controls and with no evidence UDCA-associated fetal toxicity.[122]

Fetal Nonstress Testing

There is no convincing evidence that antepartum fetal nonstress testing prevents fetal deaths from ICP, or that specific antepartum fetal tests are useful for identifying fetuses at increased risk of fetal death in this syndrome. For example, a retrospective analysis of outcomes in patients with ICP who received weekly nonstress testing (NST), daily maternal recording of fetal movement, and induction at 38 weeks or at 36 weeks found a perinatal mortality rate of 18 per 1000 (n = 328) that was not statistically significant from the control group.[11] Centers with extensive experience treating women with ICP recommend delivery when fetal maturity has been achieved to avoid late fetal deaths, even though no prospective randomized trials exist that prove superior results with this approach.[123] Electively planned early delivery shortly after the fetal lungs have matured is the accepted management for severe disease and assessment of fetal lung maturity is indicated prior to preterm induction or cesarean section.[2]

Conclusion

Cholestasis of pregnancy is an enigmatic disease whose pathogenesis is not well understood and approaches for effective management are evolving. ICP is largely a clinical diagnosis, and this disease should be considered in any woman who develops pruritus during pregnancy. The etiology of ICP is unclear, although recent data suggests a possible genetic/hereditary abnormality in the metabolism of steroids and bile acids leading to elevated maternal serum bile acids and liver enzymes and impaired ability of the fetus to transport bile acids. This results in the fetus being challenged by a greater burden of maternal bile acids than in normal pregnancies and recent studies have begun to elucidate potential mechanism by which this translates to the complications of perinatal mortality, preterm delivery, abnormal intrapartum fetal heart rate rhythm, and meconium passage.

Although symptomatically distressing for the mother due to debilitating pruritus, ICP typically follows a benign course with near total resolution of symptoms in the puerperium. Affected women may suffer from insomnia, poor weight gain due to sub-clinical steatorrhea and increased risk of postpartum hemorrhage. However, the symptoms usually disappear within a few days after delivery with rapid normalization of liver function values and serum bile acid concentrations. In contrast to the mother, the fetus may suffer devastating consequences, with perinatal mortality in untreated cases. Affected patients are treated with ursodeoxycholic acid, which can relieve pruritus and more importantly, may protect the fetus from being overwhelmed by maternal biliary constituents. Due to the increased risk of stillbirth, most experienced centers advocate delivery by the 37th week of gestation. Above all diagnostic features, a keen clinical acumen and suspicion remain as vital components for ensuring prompt diagnosis and, therefore, treatment of ICP in pregnant women presenting with pruritus.

References

1. Lammert F, Marschall HU, Glantz A et al. Intrahepatic cholestasis of pregnancy: Molecular pathogenesis, diagnosis and management. J Hepatol 2000; 33(6):1012-1021.
2. Geenes V, Williamson C. Intrahepatic cholestasis of pregnancy. World J Gastroenterol 2009; 15(17):2049-2066.
3. Joshi D, James A, Quaglia A et al. Liver disease in pregnancy. Lancet 2010; 375(9714):594-605.
4. Pusl T, Beuers U. Intrahepatic cholestasis of pregnancy. Orphanet J Rare Dis 2007; 2:26.
5. Reyes H, Gonzalez MC, Ribalta J et al. Prevalence of intrahepatic cholestasis of pregnancy in Chile. Ann Intern Med 1978; 88(4):487-493.
6. Reyes H, Taboada G, Ribalta J. Prevalence of intrahepatic cholestasis of pregnancy in La Paz, Bolivia. J Chronic Dis 1979; 32(7):499-504.
7. Reyes H. Sex hormones and bile acids in intrahepatic cholestasis of pregnancy. Hepatology 2008; 47(2):376-379.
8. Roger D, Vaillant L, Fignon A et al. Specific pruritic diseases of pregnancy. A prospective study of 3192 pregnant women. Arch Dermatol 1994; 130(6):734-739.
9. Riely CA, Bacq Y. Intrahepatic cholestasis of pregnancy. Clin Liver Dis 2004; 8(1):167-176.
10. Brites D, Rodrigues CM, Cardoso Mda C et al. Unusual case of severe cholestasis of pregnancy with early onset, improved by ursodeoxycholic acid administration. Eur J Obstet Gynecol Reprod Biol 1998; 76(2):165-168.
11. Rioseco AJ, Ivankovic MB, Manzur A et al. Intrahepatic cholestasis of pregnancy: a retrospective case-control study of perinatal outcome. Am J Obstet Gynecol 1994; 170(3):890-895.
12. Lunzer MR. Jaundice in pregnancy. Baillieres Clin Gastroenterol 1989; 3(2):467-483.
13. Reyes H, Radrigan ME, Gonzalez MC et al. Steatorrhea in patients with intrahepatic cholestasis of pregnancy. Gastroenterology 1987; 93(3):584-590.
14. Ropponen A, Sund R, Riikonen S et al. Intrahepatic cholestasis of pregnancy as an indicator of liver and biliary diseases: a population-based study. Hepatology 2006; 43(4):723-728.
15. Samsioe G, Svendsen P, Johnson P et al. Studies in cholestasis of pregnancy. V. Gallbladder disease, liver function tests, serum lipids and fatty acid composition of serum lecithin in the nonpregnant state. Acta Obstet Gynecol Scand 1975; 54(5):417-423.
16. Glantz A, Marschall HU, Mattsson LA. Intrahepatic cholestasis of pregnancy: Relationships between bile acid levels and fetal complication rates. Hepatology 2004; 40(2):467-474.
17. Reyes H. Review: intrahepatic cholestasis. A puzzling disorder of pregnancy. J Gastroenterol Hepatol 1997; 12(3):211-216.
18. Olsson R, Tysk C, Aldenborg F et al. Prolonged postpartum course of intrahepatic cholestasis of pregnancy. Gastroenterology 1993; 105(1):267-271.
19. Bacq Y, Sapey T, Brechot MC et al. Intrahepatic cholestasis of pregnancy: a French prospective study. Hepatology 1997; 26(2):358-364.
20. Reid R, Ivey KJ, Rencoret RH et al. Fetal complications of obstetric cholestasis. Br Med J 1976; 1(6014):870-872.
21. Shaw D, Frohlich J, Wittmann BA et al. A prospective study of 18 patients with cholestasis of pregnancy. Am J Obstet Gynecol 1982; 142(6 Pt 1):621-625.
22. Ammala P, Kariniemi V. Short-term variability of fetal heart rate in cholestasis of pregnancy. Am J Obstet Gynecol 1981; 141(2):217-220.

23. Laatikainen T, Tulenheimo A. Maternal serum bile acid levels and fetal distress in cholestasis of pregnancy. Int J Gynaecol Obstet 1984; 22(2):91-94.
24. Kondrackiene J, Beuers U, Zalinkevicius R et al. Predictors of premature delivery in patients with intrahepatic cholestasis of pregnancy. World J Gastroenterol 2007; 13(46):6226-6230.
25. Israel EJ, Guzman ML, Campos GA. Maximal response to oxytocin of the isolated myometrium from pregnant patients with intrahepatic cholestasis. Acta Obstet Gynecol Scand 1986; 65(6):581-582.
26. Papacleovoulou G, Abu-Hayyeh S, Williamson C. Nuclear receptor-driven alterations in bile acid and lipid metabolic pathways during gestation. Biochimica et biophysica acta 2011; 1812(8):879-887.
27. Lunzer M, Barnes P, Byth K et al. Serum bile acid concentrations during pregnancy and their relationship to obstetric cholestasis. Gastroenterology 1986; 91(4):825-829.
28. Heikkinen J. Serum bile acids in the early diagnosis of intrahepatic cholestasis of pregnancy. Obstet Gynecol 1983; 61(5):581-587.
29. Kenyon AP, Piercy CN, Girling J et al. Pruritus may precede abnormal liver function tests in pregnant women with obstetric cholestasis: a longitudinal analysis. Bjog 2001; 108(11):1190-1192.
30. Heikkinen J, Maentausta O, Ylostalo P et al. Changes in serum bile acid concentrations during normal pregnancy, in patients with intrahepatic cholestasis of pregnancy and in pregnant women with itching. Br J Obstet Gynaecol 1981; 88(3):240-245.
31. Laatikainen T, Ikonen E. Serum bile acids in cholestasis of pregnancy. Obstet Gynecol 1977; 50(3):313-318.
32. Dann AT, Kenyon AP, Seed PT et al. Glutathione S-transferase and liver function in intrahepatic cholestasis of pregnancy and pruritus gravidarum. Hepatology 2004; 40(6):1406-1414.
33. Tribe RM, Dann AT, Kenyon AP et al. Longitudinal profiles of 15 serum bile acids in patients with intrahepatic cholestasis of pregnancy. Am J Gastroenterol 2010; 105(3):585-595.
34. Bacq Y, Zarka O, Brechot JF et al. Liver function tests in normal pregnancy: a prospective study of 103 pregnant women and 103 matched controls. Hepatology 1996; 23(5):1030-1034.
35. Milkiewicz P, Gallagher R, Chambers J et al. Obstetric cholestasis with elevated gamma glutamyl transpeptidase: incidence, presentation and treatment. J Gastroenterol Hepatol 2003; 18(11):1283-1286.
36. Iber FL. Jaundice in pregnancy—a review. Am J Obstet Gynecol 1965; 91:721-753.
37. Dann AT, Kenyon AP, Wierzbicki AS et al. Plasma lipid profiles of women with intrahepatic cholestasis of pregnancy. Obstet Gynecol 2006; 107(1):106-114.
38. Sabesin SM. Cholestatic lipoproteins—their pathogenesis and significance. Gastroenterology 1982; 83(3):704-709.
39. Samsioe G, Lundborg H, Ivarsson S et al. On the occurrence of lipoprotein-x in nonhemolytic jaundice. Acta Chir Scand 1976; 142(3):187-190.
40. Carter J. Serum bile acids in normal pregnancy. Br J Obstet Gynaecol 1991; 98(6):540-543.
41. van Haelst U, Bergstein N. Electron microscopic study of the liver in so-called idiopathic jaundice of late pregnancy. Observations in six cases. Pathol Eur 1970; 5(2):198-215.
42. Eliakim M, Sadovsky E, Stein O et al. Recurrent cholestatic jaundice of pregnancy. Report of five cases and electron microscopic observations. Archives of internal medicine 1966; 117(5):696-705.
43. Ishak KG. Inherited metabolic diseases of the liver. Clin Liver Dis. May 2002;6(2):455-479, viii.
44. Morotti RA, Suchy FJ, Magid MS. Progressive familial intrahepatic cholestasis (PFIC) type 1, 2, and 3: a review of the liver pathology findings. Semin Liver Dis 2011; 31(1):3-10.
45. Kroumpouzos G, Cohen LM. Specific dermatoses of pregnancy: an evidence-based systematic review. Am J Obstet Gynecol 2003; 188(4):1083-1092.
46. Benjaminov FS, Heathcote J. Liver disease in pregnancy. Am J Gastroenterol 2004; 99(12):2479-2488.
47. Reyes H, Ribalta J, Gonzalez-Ceron M. Idiopathic cholestasis of pregnancy in a large kindred. Gut 1976; 17(9):709-713.
48. Holzbach RT, Sivak DA, Braun WE. Familial recurrent intrahepatic cholestasis of pregnancy: a genetic study providing evidence for transmission of a sex-limited, dominant trait. Gastroenterology 1983; 85(1):175-179.

49. Dalen E, Westerholm B. Occurrence of hepatic impairment in women jaundiced by oral contraceptives and in their mothers and sisters. Acta Med Scand 1974; 195(6):459-463.
50. Furhoff AK. Itching in pregnancy. A 15-year follow-up study. Acta Med Scand 1974; 196(5):403-410.
51. Reyes H, Wegmann ME, Segovia N et al. HLA in Chileans with intrahepatic cholestasis of pregnancy. Hepatology 1982; 2(4):463-466.
52. Hirvioja ML, Kivinen S. Inheritance of intrahepatic cholestasis of pregnancy in one kindred. Clin Genet 1993; 43(6):315-317.
53. Mella JG, Roschmann E, Glasinovic JC et al. Exploring the genetic role of the HLA-DPB1 locus in Chileans with intrahepatic cholestasis of pregnancy. J Hepatol 1996; 24(3):320-323.
54. Savander M, Ropponen A, Avela K et al. Genetic evidence of heterogeneity in intrahepatic cholestasis of pregnancy. Gut 2003; 52(7):1025-1029.
55. Elferink RO. Cholestasis. Gut 2003; 52 Suppl 2:ii42-ii48.
56. Davit-Spraul A, Gonzales E, Baussan C et al. Progressive familial intrahepatic cholestasis. Orphanet J Rare Dis 2009; 4:1.
57. Bull LN, van Eijk MJ, Pawlikowska L et al. A gene encoding a P-type ATPase mutated in two forms of hereditary cholestasis. Nat Genet 1998; 18(3):219-224.
58. Klomp LW, Vargas JC, van Mil SW et al. Characterization of mutations in ATP8B1 associated with hereditary cholestasis. Hepatology 2004; 40(1):27-38.
59. Strautnieks SS, Bull LN, Knisely AS et al. A gene encoding a liver-specific ABC transporter is mutated in progressive familial intrahepatic cholestasis. Nat Genet 1998; 20(3):233-238.
60. van Mil SW, van der Woerd WL, van der Brugge G et al. Benign recurrent intrahepatic cholestasis type 2 is caused by mutations in ABCB11. Gastroenterology 2004; 127(2):379-384.
61. Jacquemin E, De Vree JM, Cresteil D et al. The wide spectrum of multidrug resistance 3 deficiency: from neonatal cholestasis to cirrhosis of adulthood. Gastroenterology 2001; 120(6):1448-1458.
62. Oude Elferink RP, Paulusma CC, Groen AK. Hepatocanalicular transport defects: pathophysiologic mechanisms of rare diseases. Gastroenterology 2006; 130(3):908-925.
63. de Pagter AG, van Berge Henegouwen GP, ten Bokkel Huinink JA et al. Familial benign recurrent intrahepatic cholestasis. Interrelation with intrahepatic cholestasis of pregnancy and from oral contraceptives? Gastroenterology 1976; 71(2):202-207.
64. Whitington PF, Freese DK, Alonso EM et al. Clinical and biochemical findings in progressive familial intrahepatic cholestasis. J Pediatr Gastroenterol Nutr 1994; 18(2):134-141.
65. Jacquemin E, Cresteil D, Manouvrier S et al. Heterozygous nonsense mutation of the MDR3 gene in familial intrahepatic cholestasis of pregnancy. Lancet 1999; 353(9148):210-211.
66. Dixon PH, Weerasekera N, Linton KJ et al. Heterozygous MDR3 missense mutation associated with intrahepatic cholestasis of pregnancy: evidence for a defect in protein trafficking. Hum Mol Genet 2000; 9(8):1209-1217.
67. Keitel V, Vogt C, Haussinger D et al. Combined mutations of canalicular transporter proteins cause severe intrahepatic cholestasis of pregnancy. Gastroenterology 2006; 131(2):624-629.
68. Lucena JF, Herrero JI, Quiroga J et al. A multidrug resistance 3 gene mutation causing cholelithiasis, cholestasis of pregnancy, and adulthood biliary cirrhosis. Gastroenterology 2003; 124(4):1037-1042.
69. Mullenbach R, Linton KJ, Wiltshire S et al. ABCB4 gene sequence variation in women with intrahepatic cholestasis of pregnancy. J Med Genet 2003; 40(5):e70.
70. Gendrot C, Bacq Y, Brechot MC et al. A second heterozygous MDR3 nonsense mutation associated with intrahepatic cholestasis of pregnancy. J Med Genet 2003; 40(3):e32.
71. Eloranta ML, Heiskanen JT, Hiltunen MJ et al. Multidrug resistance 3 gene mutation 1712delT and estrogen receptor alpha gene polymorphisms in Finnish women with obstetric cholestasis. Eur J Obstet Gynecol Reprod Biol 2002; 105(2):132-135.
72. Painter JN, Savander M, Sistonen P et al. A known polymorphism in the bile salt export pump gene is not a risk allele for intrahepatic cholestasis of pregnancy. Scand J Gastroenterol 2004; 39(7):694-695.
73. Pauli-Magnus C, Lang T, Meier Y et al. Sequence analysis of bile salt export pump (ABCB11) and multidrug resistance p-glycoprotein 3 (ABCB4, MDR3) in patients with intrahepatic cholestasis of pregnancy. Pharmacogenetics 2004; 14(2):91-102.

74. Eloranta ML, Hakli T, Hiltunen M et al. Association of single nucleotide polymorphisms of the bile salt export pump gene with intrahepatic cholestasis of pregnancy. Scand J Gastroenterol 2003; 38(6):648-652.
75. Mullenbach R, Bennett A, Tetlow N et al. ATP8B1 mutations in British cases with intrahepatic cholestasis of pregnancy. Gut 2005; 54(6):829-834.
76. Dixon PH, van Mil SW, Chambers J et al. Contribution of variant alleles of ABCB11 to susceptibility to intrahepatic cholestasis of pregnancy. Gut 2009; 58(4):537-544.
77. Painter JN, Savander M, Ropponen A et al. Sequence variation in the ATP8B1 gene and intrahepatic cholestasis of pregnancy. Eur J Hum Genet 2005; 13(4):435-439.
78. Meier Y, Zodan T, Lang C et al. Increased susceptibility for intrahepatic cholestasis of pregnancy and contraceptive-induced cholestasis in carriers of the 1331T>C polymorphism in the bile salt export pump. World J Gastroenterol 2008; 14(1):38-45.
79. Sookoian S, Castano G, Burgueno A et al. Association of the multidrug-resistance-associated protein gene (ABCC2) variants with intrahepatic cholestasis of pregnancy. J Hepatol 2008; 48(1):125-132.
80. Van Mil SW, Milona A, Dixon PH et al. Functional variants of the central bile acid sensor FXR identified in intrahepatic cholestasis of pregnancy. Gastroenterology 2007; 133(2):507-516.
81. Paulusma CC, Kool M, Bosma PJ et al. A mutation in the human canalicular multispecific organic anion transporter gene causes the Dubin-Johnson syndrome. Hepatology 1997; 25(6):1539-1542.
82. Rastogi A, Krishnani N, Pandey R. Dubin-Johnson syndrome—a clinicopathologic study of twenty cases. Indian J Pathol Microbiol 2006; 49(4):500-504.
83. Seligsohn U, Shani M. The Dubin Johnson syndrome and pregnancy. Acta Hepatogastroenterol (Stuttg) 1977; 24(3):167-169.
84. Reyes H, Sjovall J. Bile acids and progesterone metabolites in intrahepatic cholestasis of pregnancy. Ann Med 2000; 32(2):94-106.
85. Kreek MJ. Female sex steroids and cholestasis. Semin Liver Dis 1987; 7(1):8-23.
86. Gonzalez MC, Reyes H, Arrese M et al. Intrahepatic cholestasis of pregnancy in twin pregnancies. J Hepatol 1989; 9(1):84-90.
87. Williamson C, Hems LM, Goulis DG et al. Clinical outcome in a series of cases of obstetric cholestasis identified via a patient support group. Bjog 2004; 111(7):676-681.
88. Trauner M, Arrese M, Soroka CJ et al. The rat canalicular conjugate export pump (Mrp2) is down-regulated in intrahepatic and obstructive cholestasis. Gastroenterology 1997; 113(1):255-264.
89. Stieger B, Fattinger K, Madon J et al. Drug- and estrogen-induced cholestasis through inhibition of the hepatocellular bile salt export pump (Bsep) of rat liver. Gastroenterology 2000; 118(2):422-430.
90. Giusti G, Piccinino F, Ricciardi I et al. Abnormal steroid sulfate in plasma of women with intrahepatic cholestasis of pregnancy. Acta Hepatogastroenterol (Stuttg) 1979; 26(3):203-206.
91. Meng LJ, Reyes H, Palma J et al. Profiles of bile acids and progesterone metabolites in the urine and serum of women with intrahepatic cholestasis of pregnancy. J Hepatol 1997; 27(2):346-357.
92. Meng LJ, Reyes H, Palma J et al. Effects of ursodeoxycholic acid on conjugated bile acids and progesterone metabolites in serum and urine of patients with intrahepatic cholestasis of pregnancy. J Hepatol 1997; 27(6):1029-1040.
93. Anakk S, Watanabe M, Ochsner SA et al. Combined deletion of Fxr and Shp in mice induces Cyp17a1 and results in juvenile onset cholestasis. J Clin Invest 2011; 121(1):86-95.
94. Khurana S, Raufman JP, Pallone TL. Bile acids regulate cardiovascular function. Clin Transl Sci 2011; 4(3):210-218.
95. Desai MS, Shabier Z, Taylor M et al. Hypertrophic cardiomyopathy and dysregulation of cardiac energetics in a mouse model of biliary fibrosis. Hepatology 2010; 51(6):2097-2107.
96. Sheikh Abdul Kadir SH, Miragoli M, Abu-Hayyeh S et al. Bile acid-induced arrhythmia is mediated by muscarinic M2 receptors in neonatal rat cardiomyocytes. PLoS One 2010; 5(3):e9689.
97. Miragoli M, Kadir SH, Sheppard MN et al. A protective antiarrhythmic role of ursodeoxycholic acid in an in vitro rat model of the cholestatic fetal heart. Hepatology 2011; 54(4):1282-1292.
98. Gonzalez MC, Iglesias J, Tiribelli C et al. Epomediol ameliorates pruritus in patients with intrahepatic cholestasis of pregnancy. J Hepatol 1992; 16(1-2):241-242.

99. Kremer AE, Martens JJ, Kulik W et al. Lysophosphatidic acid is a potential mediator of cholestatic pruritus. Gastroenterology 2010; 139(3):1008-1018, 1018 e1001.
100. Oude Elferink RP, Kremer AE, Beuers U. Mediators of pruritus during cholestasis. Curr Opin Gastroenterol 2011; 27(3):289-293.
101. Parrill AL, Baker DL. Autotaxin inhibitors: a perspective on initial medicinal chemistry efforts. Expert Opin Ther Pat 2010; 20(12):1619-1625.
102. EASL Clinical Practice Guidelines: management of cholestatic liver diseases. J Hepatol 2009; 51(2):237-267.
103. Bergasa NV. The pruritus of cholestasis. Semin Dermatol 1995; 14(4):302-312.
104. Sadler LC, Lane M, North R. Severe fetal intracranial haemorrhage during treatment with cholestyramine for intrahepatic cholestasis of pregnancy. Br J Obstet Gynaecol 1995; 102(2):169-170.
105. Frezza M, Pozzato G, Chiesa L et al. Reversal of intrahepatic cholestasis of pregnancy in women after high dose S-adenosyl-L-methionine administration. Hepatology 1984; 4(2):274-278.
106. Frezza M, Centini G, Cammareri G et al. S-adenosylmethionine for the treatment of intrahepatic cholestasis of pregnancy. Results of a controlled clinical trial. Hepatogastroenterology 1990; 37 Suppl 2:122-125.
107. Ribalta J, Reyes H, Gonzalez MC et al. S-adenosyl-L-methionine in the treatment of patients with intrahepatic cholestasis of pregnancy: a randomized, double-blind, placebo-controlled study with negative results. Hepatology 1991; 13(6):1084-1089.
108. Talwalkar JA, Lindor KD. Primary biliary cirrhosis. Lancet 2003; 362(9377):53-61.
109. Paumgartner G, Beuers U. Ursodeoxycholic acid in cholestatic liver disease: Mechanisms of action and therapeutic use revisited. Hepatology 2002; 36(3):525-531.
110. Beuers U. Drug insight: Mechanisms and sites of action of ursodeoxycholic acid in cholestasis. Nat Clin Pract Gastroenterol Hepatol 2006; 3(6):318-328.
111. Palma J, Reyes H, Ribalta J et al. Ursodeoxycholic acid in the treatment of cholestasis of pregnancy: a randomized, double-blind study controlled with placebo. J Hepatol 1997; 27(6):1022-1028.
112. Diaferia A, Nicastri PL, Tartagni M et al. Ursodeoxycholic acid therapy in pregnant women with cholestasis. Int J Gynaecol Obstet 1996; 52(2):133-140.
113. Roncaglia N, Locatelli A, Arreghini A et al. A randomised controlled trial of ursodeoxycholic acid and S-adenosyl-l-methionine in the treatment of gestational cholestasis. Bjog 2004; 111(1):17-21.
114. Kondrackiene J, Beuers U, Kupcinskas L. Efficacy and safety of ursodeoxycholic acid versus cholestyramine in intrahepatic cholestasis of pregnancy. Gastroenterology 2005; 129(3):894-901.
115. Glantz A, Marschall HU, Lammert F et al. Intrahepatic cholestasis of pregnancy: a randomized controlled trial comparing dexamethasone and ursodeoxycholic acid. Hepatology 2005; 42(6):1399-1405.
116. Nicastri PL, Diaferia A, Tartagni M et al. A randomised placebo-controlled trial of ursodeoxycholic acid and S-adenosylmethionine in the treatment of intrahepatic cholestasis of pregnancy. Br J Obstet Gynaecol 1998; 105(11):1205-1207.
117. Binder T, Salaj P, Zima T et al. Randomized prospective comparative study of ursodeoxycholic acid and S-adenosyl-L-methionine in the treatment of intrahepatic cholestasis of pregnancy. J Perinat Med 2006; 34(5):383-391.
118. Davies MH, da Silva RC, Jones SR et al. Fetal mortality associated with cholestasis of pregnancy and the potential benefit of therapy with ursodeoxycholic acid. Gut 1995; 37(4):580-584.
119. Serrano MA, Brites D, Larena MG et al. Beneficial effect of ursodeoxycholic acid on alterations induced by cholestasis of pregnancy in bile acid transport across the human placenta. J Hepatol 1998; 28(5):829-839.
120. Brites D, Rodrigues CM, Oliveira N et al. Correction of maternal serum bile acid profile during ursodeoxycholic acid therapy in cholestasis of pregnancy. J Hepatol 1998; 28(1):91-98.
121. Zimber A, Zusman I. Effects of secondary bile acids on the intrauterine development in rats. Teratology 1990; 42(3):215-224.
122. Zapata R, Sandoval L, Palma J et al. Ursodeoxycholic acid in the treatment of intrahepatic cholestasis of pregnancy. A 12-year experience. Liver Int 2005; 25(3):548-554.
123. Heinonen S, Kirkinen P. Pregnancy outcome with intrahepatic cholestasis. Obstet Gynecol 1999; 94(2):189-193.

Chapter 6

Preeclampsia-Induced Liver Disease and HELLP Syndrome

Dilip Bearelly, Ghassan M. Hammoud, Gretchen Koontz, David C. Merrill and Jamal A. Ibdah*

Abstract

Hypertensive disorders of pregnancy remain a devastating disease for both the mother and the fetus. Preeclampsia and HELLP syndrome, two disorders unique to pregnancy, remain a major cause of maternal and neonatal mortality and morbidity worldwide. Unfortunately, the diagnosis is not always straightforward as patients may present with varying symptoms and degrees of severity. The laboratory evaluation remains the main diagnostic modality.

In both of these disorders, the liver is a major target with often devastating consequences. The pathogenesis of hepatic damage in cases of severe preeclampsia and HELLP syndrome is not well understood. The pathogenesis of liver damage is difficult to study as liver enzyme evaluations and maternal symptoms often correlate poorly with pathological findings. The characteristic liver pathology seen is that of dense fibrin accumulation leading to hemorrhage, necrosis and wide areas of liver infarction. The only treatment available remains to be expeditious delivery. Corticosteroids may be of potential benefit, although this remains unclear. Other specific treatment option such as liver transplantation is performed if hepatic failure ensues. Hepatic artery embolization is another option in the setting of liver rupture.

Despite extensive research and advances in technology and antepartum care, there still remain many unanswered questions. The potential molecular mechanisms mitigating the liver involvement in this disease are discussed, but at present are only speculative. Areas of research continue to focus on the etiology of the disease and markers of disease progression. Future research will likely focus on the genetic risk and etiology of the disease.

Introduction

Preeclampsia is a syndrome that is characterized by heterogeneous clinical and laboratory findings and considered a disorder unique to pregnancy. The clinical findings of preeclampsia can manifest as a maternal syndrome (hypertension, proteinuria, and/or various symptoms) and/or a fetal syndrome (growth restriction). Preeclampsia is a major cause of maternal and neonatal mortality and morbidity worldwide. It is the most common medical disorder complicating pregnancy, affecting 7 to 10% of all women. As much as 15 to 20% of maternal mortality in developed countries can be attributed to preeclampsia.[1,2] Not every patient with preeclampsia may exhibit all of the three features. Several studies identified that the syndrome may present

*Corresponding Author: Jamal A. Ibdah—Department of Internal Medicine, Division of Gastroenterology and Hepatology, University of Missouri, Columbia, Missouri, USA. Email: ibdahj@health.missouri.edu

with only two (and not three) of its components. Pregnant women with preeclampsia may present with involvement of other organs such as liver, brain, kidneys, etc. Liver involvement in preeclampsia is not common but when present signifies severe disease. Hepatic involvement manifested by elevation in serum aminotransferases and is known to occur in up to 10% of cases of severe preeclampsia.[2,3] Preeclampsia-induced liver disease is a disorder unique to pregnancy and is frequently seen in the third trimester. Severe preeclampsia is defined by extreme elevations in systemic blood pressure and evidence of organ compromise. Severity ranges from a mild disorder to a life threatening disorder marked by seizures, laboratory abnormalities, and fetal compromise. Preeclampsia with seizures is referred to as eclampsia.

For many years, hemolysis, abnormal liver function tests, and thrombocytopenia have been recognized as complications of preeclampsia-eclampsia. In fact, some of these components of the disease have been reported in the obstetrical literature for over a century. It was not until 1982, that a syndrome separate from severe preeclampsia was proposed by Weinstein by adding a specific set of criteria to the diagnosis.[4] This disorder was termed HELLP syndrome, with (H) for hemolysis, (EL) for elevated liver functions tests, and (LP) for low platelet counts (Table 1). In his original paper, Weinstein described the severity of this disease. He studied 29 patients over a period of 30 months. The reported perinatal death rate at that time was 9.4% (3/32) and the maternal mortality rate was 3% (1/32). Weinstein concluded that the entity was more common than once thought and mandated aggressive therapy.

The syndrome of hemolysis, elevated liver enzymes, and low platelets (HELLP) is now recognized as a variant of severe preeclampsia; and a disease unique to pregnancy; which carries statistically significant perinatal risks to both the mother and the fetus. Table 2 lists liver disorders unique to pregnancy throughout the gestation period.

Table 1. Main diagnostic criteria of HELLP syndrome

Tennessee Classification*	Mississippi Classification*
Complete syndrome:	Class 1: platelets $\leq 50 \times 10^9$/L
Platelets $\leq 100 \times 10^9$/L	AST or ALT $\geq$ 70 units/L
AST $\geq$ 70 units/L	LDH $\geq$ 600 units/L
LDH $\geq$ 600 units/L	
Incomplete syndrome:	Class 2: platelets $\leq 100 \times 10^9$/L
Any one or two of the above	$\geq 50 \times 10^9$/L
	AST or ALT $\geq$ 70 units/L
	LDH $\geq$ 600 units/L
	Class 3: platelets $\leq 150 \times 10^9$/L
	$\geq 100 \times 10^9$/L
	AST or ALT $\geq$ 40 units/L
	LDH $\geq$ 600 units/L

*Adapted from Audibert F, Friedman SA, Frangieh AY et al. Clinical utility of strict diagnostic criteria for the HELLP (hemolysis, elevated liver enzymes, and low platelets) syndrome. Am J Obstet Gynecol 1996; 175:460-464;[20] and from Martin JN Jr, Rinehart BK, May WL et al. The spectrum of severe preeclampsia: comparative analysis by HELLP (hemolysis, elevated liver enzyme levels, and low platelet count) syndrome classification. Am J Obstet Gynecol 1999; 180(6 Pt 1):1373-1384.[22]
ALT: alanine aminotransferase; AST: aspartate aminotransferase; HELLP: hemolysis, elevated liver enzymes, and low platelet count; LDH: lactate dehydrogenase.

Table 2. Liver disease in pregnancy

<table>
<tr><th>1st Trimester</th><th>2nd Trimester</th><th>3rd Trimester</th><th>Post Partum</th><th>Prevalence</th></tr>
<tr><td colspan="5">Associated with preeclampsia</td></tr>
<tr><td></td><td colspan="3">HELLP syndrome</td><td>0.2-0.6%</td></tr>
<tr><td></td><td></td><td>Acute fatty liver</td><td></td><td>0.005-0.010%</td></tr>
<tr><td></td><td colspan="2">Severe preeclampsia</td><td></td><td>5-7%</td></tr>
<tr><td colspan="5">Not associated/preeclampsia</td></tr>
<tr><td>Hyperemesis gravidarum</td><td></td><td></td><td></td><td>0.3-1%</td></tr>
<tr><td></td><td colspan="3">Intrahepatic cholestasis of pregnancy</td><td>0.1-0.3%</td></tr>
</table>

Terminology

The reported incidence of HELLP syndrome ranges from 2 to 12% of patients with preeclampsia, reflecting the different definitions and criteria used in the diagnosis. In a review by Sibai and associates, it was shown that there is considerable difference concerning the terminology, incidence, cause, diagnosis, and management of HELLP syndrome.[5] Weinstein considered HELLP syndrome to be a "unique variant" of preeclampsia whereas others have considered it to be a misdiagnosis of preeclampsia.[4] Still others consider HELLP syndrome to be a separate disease entity entirely or a variant of mild disseminated intravascular coagulation (DIC).

Epidemiology

Preeclampsia is the most common medical disorder complicating pregnancy, affecting 7 to 10% of all women.[1] As much as 15 to 20% of maternal mortality in developed countries can be attributed to preeclampsia. HELLP syndrome complicates 2 to 20% of cases with severe preeclampsia and about 0.2 to 0.6% of all pregnancies. In a large prospective study of 442 patients with preeclampsia, the incidence of HELLP syndrome was 20%.[6] In patients with eclampsia, the incidence of HELLP syndrome was found to be 10% in one study and 30% in another study. In a secondary analysis of information collected in the ECLAXIR study in France between May 2003 and October 2006, the data from 284 white European, 84 Maghrebian and 158 African women were evaluated in a case-control study of the genetic and endothelial determinants of preeclampsia.[7] This study suggests that ethnic origin may have an effect on the severity of the preeclampsia. Considerable heterogeneity in pregnancy outcomes is evident depending on gestational age at onset of preeclampsia.

Clinical Presentation

Hypertensive disorders of pregnancy are more common in the extreme maternal age ranges. Patients with preeclampsia-eclampsia and HELLP syndrome may present with various signs and symptoms, none of which are diagnostic. Pregnant women usually present in their third trimester with complaints of malaise (90%), epigastric or right upper-quadrant pain (90%), nausea or vomiting (50%), or nonspecific viral-like symptoms.[8] Although, majority of these patients present in the third trimester, it is not uncommon to see these cases in the later part of second trimester, or in the postpartum period.[6] These clinical features are reported in various studies and case reports with varying frequency but are seen in at least 50% of cases. However, these clinical features are also common presentation for several other benign and serious pregnancy and nonpregnancy related conditions (Table 3). For this reason, pregnant women with

Table 3. Differential diagnosis

Hepatic	
	Acute fatty liver of pregnancy
	Intrahepatic cholestasis of pregnancy
	Cholecystitis
	Cholangitis
	Viral hepatitis
	Acute pancreatitis
	Gastritis
	Gastric ulcer
Non-Hepatic	
	Benign thrombocytopenia of pregnancy
	Thrombotic thrombocytopenic purpura (TTP)
	Hemolytic uremic syndrome (HUS)
	Ideopathic thrombocytopenic purpura (ITP)
	Antiphospholipid antibody syndrome
	Folate deficiency
	Systemic lupus erythematosus (SLE)
	Septic or hemorrhagic shock

any concerning symptoms should undergo a diagnostic work-up including a complete blood count, platelet count, liver evaluation, and urine dipstick for protein irrespective of their blood pressure.[9] Presence of abnormal urine dipstick for protein should be followed by quantitative evaluation for protein in a 24 hour urine specimen.

Typically pregnant women present in their third trimester with systemic hypertension, proteinuria, and peripheral edema. Patients may present with increased weight gain, headache and visual disturbances, and gastric complaints of nausea and vomiting. The severity of preeclampsia is based on the presence of cerebral disturbances, proteinuria greater than 5 g/24 hours, or evidence of thrombocytopenia or hemolysis. Generally, proteinuria is defined as equal to or greater than 300mg of protein in an adequate 24 hour specimen. Thrombocytopenia and hemolysis may or may not be present. Hepatic involvement occurs in approximately 10% of severe preeclampsia cases.[1] It is not uncommon that these women may have transient liver enzyme elevations without clinical signs of pain.

Abdominal pain is common and may be present in about 50% of the patients. Abdominal pain is usually encountered in the right upper quadrant, epigastric or substernal region and often associated with laboratory abnormalities defining HELLP syndrome. Abdominal pain is generally absent in other disorders unique to pregnancy such as cholestasis of pregnancy and hyperemesis of pregnancy; however it is frequently encountered in HELLP and AFLP (Table 4). Although HELLP syndrome may have symptoms similar to preeclampsia and is one of the criteria that can define severe preeclampsia, it can develop in women who might not have any other signs or symptoms of preeclampsia. Preeclampsia is not a prerequisite for HELLP syndrome and hypertension, if present, does not have to be severe. Severe hypertension defined as systolic blood pressure ≥160 mm Hg and diastolic blood pressure ≥110 mm Hg, is not a constant or even a frequent finding in HELLP syndrome. In the initial case series published

Table 4. Liver disease associated with preeclampsia: key points

Severe Preeclampsia and Eclampsia

Presentation: after week 22

Prevalence: 5 to 7%; higher in multiple gestation

Symptoms: high blood pressure; proteinuria; edema; seizure; renal failure; pulmonary edema

Laboratory features: platelets > 70,000; urine protein >5 gm/24 hrs; abnormal Liver enzymes (10%);

Treatment: blood pressure control, beta-blockers, methyldopa, magnesium sulfate for eclampsia, early delivery

Outcome: maternal death rate 1%

HELLP Syndrome

Presentation: later part of second trimester or third trimester or immediate postpartum period

Prevalence: 0.1% of all pregnancies

Symptoms: epigastric or RUQ pain, nausea and vomiting, overlap with signs and symptoms of preeclampsia

Laboratory features: platelets < 100,000; hemolysis; abnormal liver enzymes where AST and ALT levels may be >1,000 U/L; prothrombin time may remain normal; normal fibrinogen

Treatment: prompt delivery

Outcome: maternal death rate 5%; hepatic rupture in 1%; fetal death rate 1 to 30%

Acute Fatty Liver of Pregnancy

Presentation: during third trimester; 50% of patients may have eclampsia

Prevalence: 0.01% of all pregnancies; higher prevalence in multiple gestations, primiparous women, male fetus

Symptoms: nausea, vomiting, abdominal pain, jaundice; can progress rapidly to hepatic failure, hypoglycemia

Laboratory features: platelets < 100,000 (normal 150-450 × 10^9/L); AST and ALT 300-1,000 U/L; decreased antithrombin III; elevated prothrombin time; low fibrinogen; elevated bilirubin; disseminated intravascular coagulation

Treatment: prompt delivery; liver transplant

Outcome: maternal death rate ≤ 10%; fetal death rate up to 45%

by Weinstein et al, 13 of the 29 patients had had an admission blood pressure of 160/110 mm Hg or greater.[4] In another study (n = 27), in addition to hemolysis, liver enzyme elevation, and thrombocytopenia as described by Weinstein, all patients had pregnancy induced hypertension but only 66% of the 18 primigravidas and 44% of the nine multigravidas had severe hypertension on admission.[10] In another case series none of the 6 patients had blood pressure greater than 140/90 or proteinuria.[11]

Hemolysis, defined as the presence of microangiopathic hemolytic anemia, is the hallmark of the triad of HELLP syndrome.[8] The classical findings of microangiopathic hemolysis include significant drop in hemoglobin levels, elevated serum indirect bilirubin, low serum haptoglobin levels, elevated lactate dehydrogenase (LDH) levels and abnormal peripheral smear (schistocytes, burr cells, echinocytes).[12-18] Several published reports included patients who had no documentation of hemolysis; hence, these patients did not quite fit the criteria for HELLP syndrome but may fit the criteria for "ELLP" syndrome. Even in studies where hemolysis was mentioned, the diagnosis was based on the presence of abnormal peripheral smear without any description of type or degree of abnormalities or elevated LDH levels.[10,19]

The same ambiguity exists with the use of abnormal liver function tests for defining HELLP syndrome. There is no consensus regarding the degree of liver enzyme elevation that is used for diagnosing HELLP syndrome in the published literature. In a review, Sibai et al make a specific mention of the criteria used to define abnormal liver function tests.[8] They were either discussed as abnormal in some studies or indicated as values that ranged from 17 to 72 U/L for AST and ALT. These values suggest that these women did not have HELLP syndrome but may have had either severe preeclampsia with thrombocytopenia, low platelet syndrome, gestational thrombocytopenia, or immune thrombocytopenic purpura.[8]

Low platelet count is another abnormality required to make a diagnosis of HELLP syndrome. However, there are no defining criteria for low platelet count. Published studies used a wide range of cut off values. Just as discussed above in relation to other criteria for HELLP syndrome, some of these patients may have had severe preeclampsia and not HELLP syndrome.

Adding to this confusion, HELLP syndrome shares many clinical and laboratory characteristics with equally serious conditions like systemic inflammatory response syndrome (SIRS), disseminated intravascular coagulation (DIC), thrombotic thrombocytopenic purpura (TTP), hemolytic uremic syndrome (HUS) and acute fatty liver of pregnancy (AFLP) and can be easily confused with these conditions. Diagnosis of HELLP syndrome is heavily relied upon laboratory investigation. Hence the differential diagnosis for this condition should include a varying consideration including hepatic, hematological and other systemic conditions. In trying to differentiate it from acute fatty liver of pregnancy which occurs in the third trimester of pregnancy, a common clinical observation is that the liver dysfunction is more pronounced in the later with coagulopathy, hypoglycemia and renal failure are present. The coagulopathy in AFLP is due to acute liver failure, whereas in HELLP syndrome coagulopathy develops as a part of the DIC syndrome with microangiopathic hemolytic anemia.

Women with a history of previous preeclampsia are at increased risk of preeclampsia and other adverse pregnancy outcomes in subsequent pregnancies. The magnitude of this risk is dependent on gestational age at time of disease onset, severity of disease, and presence or absence of pre-existing medical disorders. Recent studies have confirmed that there is no single biomarker that can be clinically useful for the prediction of recurrent preeclampsia.

Diagnosis

Two major diagnostic classification systems are currently used for the classification of HELLP syndrome (Table 1). In the Tennessee classification system,[20] a diagnosis of the complete form of HELLP syndrome requires the presence of all three major components, whereas partial or incomplete HELLP syndrome consists of only one or two elements of the triad.[21,22] The presence of an abnormal peripheral smear (e.g., microangioplastic anemia with schistocytosis), thrombocytopenia, and elevated levels of AST, ALT, bilirubin, and lactate dehydrogenase (LDH) is diagnostic.[23] The Mississippi classification system (Table 1) has been proposed for assessment of the severity of the pathologic process, with class 1 HELLP syndrome having a worse prognosis and longer hospital stay than either class 2 or class 3. This classification system is based on the degree of thrombocytopenia and the extent of elevation in transaminase and LDH levels, as shown in Table 1. The platelet count and serum LDH levels are found not only to be moderately predictive of the severity of the disease but also to indicate the speed of recovery. Because of an initial nonspecific presentation, HELLP syndrome can be confused with acute viral hepatitis, hemolytic uremic syndrome (HUS), thrombotic thrombocytopenic purpura (TTP), antiphospholipid syndrome, and acute fatty liver of pregnancy (AFLP). Both HELLP syndrome and acute fatty liver of pregnancy occur in the third trimester and have similar presentations, but liver dysfunction is usually more pronounced in the latter and is more frequently associated with coagulopathy, hypoglycemia, and renal failure (Table 4). The coagulopathy of AFLP is due to liver failure, whereas in HELLP syndrome coagulopathy develops as a part of disseminated intravascular coagulation (DIC) syndrome. There is no consensus in the literature regarding which laboratory values should be used in the diagnosis of HELLP. Weinstein first described the laboratory abnormalities of HELLP, but did not indicate whether it was necessary to obtain certain concentrations of bilirubin, serum aspartate transaminase (AST), or serum alanine transaminase (ALT) before reaching a diagnosis.[4] Martin and coworkers, in a retrospective review of 302 cases of HELLP syndrome at the University of Mississippi, Jackson classified cases based on platelet count nadir (Table 1):[24]

Class 1: a platelet nadir below 50,000/mm3
Class 2: a platelet nadir between 51,000 and 100,000/mm3
Class 3: a platelet nadir between 101,000 and 150,000/mm3

These classes have been used to predict the postpartum disease recovery, risk of recurrence of HELLP syndrome, perinatal outcome, and the need for plasmapheresis.[5]

There is no consensus regarding the absolute values of laboratory abnormalities needed to make the diagnosis of HELLP syndrome.

Pathological Findings

The pathogenesis of hepatic damage in cases of severe preeclampsia and HELLP syndrome, in particular, is not well understood. The natural progression of the disease is difficult to study as the known treatment, delivery, is almost always rapidly undertaken. Much of what is known regarding the progression of the liver disease emanates from an important review by Rolfes et al.[25] This review of 102 cases submitted to the Armed Forces Institute of Pathology between 1920 and 1984, included women with preeclampsia, eclampsia and unclassified toxemia. Cases were selected if the clinical documentation indicated a firm diagnosis by the obstetrician of toxemia, preeclampsia, or eclampsia and there was evidence of proteinuria after the 20th week of gestation. A total of 102 liver biopsies were reviewed including the full autopsy on 97 patients. The age and race of the mothers corresponded to that of the normal obstetric population. The onset of disease ranged from 20-40 weeks gestation and symptoms associated with liver dysfunction included right upper quadrant and epigastric pain often accompanied by nausea and vomiting. Jaundice was seen in 40% of patients.

Most of the maternal deaths were attributed to central nervous system catastrophes, including large cerebral and brain stem hemorrhages, extensive thrombosis and infarction, and severe cerebral edema with brain herniation. Liver disease contributed to the mortality in 17 cases in which the brains were either normal or only mildly edematous at autopsy. Prior to death, hepatic failure had been clinically diagnosed in only four. Laboratory evaluations were available in 26 women, all demonstrated serum transaminase elevations ranging from 350-3720 U/L, most above 500 U/L. Bilirubin elevations above the normal physiologic values of pregnancy were seen, with the most marked increases occurring in patients with extensive hepatic infarction. Two of these women developed hematomas of the right lobe, one of which ruptured Glisson's capsule resulting in an exsanguinating hemorrhage despite surgical intervention. These women often developed other complications including shock, pulmonary edema, gastrointestinal bleeding and renal failure.

Microangiopathic hemolytic anemia is the hallmark of HELLP syndrome. It is a result of the passage of red blood cells through small blood vessels with damaged intima and fibrin deposition. The classic hepatic lesion associated with the HELLP syndrome is periportal or focal parenchymal necrosis in which hyaline deposits of fibrin-like material can be seen in the sinusoids.[26] On review of the autopsies and liver biopsies by Rolfes the characteristic liver abnormality seen was extensive periportal lesions that produced widespread parenchymal hemorrhage and necrosis.[25]

In addition, there were large areas of infarction involving 50-90% of the liver. Histologically, progressive accumulation of fibrin was seen with dense wide bands of fibrin eventually replacing all adjacent liver cells. Fibrin deposition in the portal tracts accompanied sinusoidal deposition. Mild extravasation of red blood cells was often seen in these areas. Thrombi in the capillaries were most frequently seen, less frequently hepatic arterial thrombi and rarely intrahepatic portal veins were occluded. Scanning electron microscopy on five specimens confirmed these findings. Most often, in areas of portal and parenchymal hemorrhage, fibrin strands were absent. However, in 31 cases there was evidence of both fibrin deposition and hemorrhage. Most often, dense deposits of fibrin developed in the immediate periportal regions while more peripherally the parenchyma was effaced by hemorrhage. Resolution of these changes was seen in three patients who had died more than two weeks after the onset of the disease. This was an exceptional review, giving insight into the pathologic progression of liver disease in pregnancies complicated by preeclampsia and HELLP syndrome.[25] Such information is rarely identified on liver biopsies in nonfatal cases.

A small study by Aarnoudse in 1986, found a correlation between patient symptoms and liver enzyme abnormalities with periportal and/or focal parenchymal liver lesions.[11] Whether these pathological findings correlate with abnormal laboratory values and the severity of patients' symptoms, however, remains unclear. Barton et al reviewed liver biopsies on eleven patients with clinically diagnosed HELLP syndrome.[27] Using immunofluorescence, he found fibrin microthrombi and fibrinogen deposits in the sinusoids in areas of hepatocellular necrosis and in sinusoids of histologically normal parenchyma.[27] There was no correlation between laboratory abnormalities and the underlying histopathologic findings. However, there was a trend in that those with severe preeclampsia had serum transaminase levels greater than 500 U/L. It was not statistically significant due to the small study size of eleven patients. He concluded that all patients with clinically defined HELLP syndrome should be treated aggressively regardless of the degree of their laboratory abnormalities.

Radiological Investigation

The diagnosis of HELLP syndrome usually rests on clinical findings and laboratory investigation including biological background and test results. Imaging studies like ultrasound, computed tomography and magnetic resonance imaging are rarely helpful in making the diagnosis of HELLP syndrome. However, they may be useful to evaluate for other differential diagnosis considered in this clinical presentation. Imaging studies have a larger utility in diagnosing complications of hepatic infarct, hematoma, and rupture.[28] Given the risk of radiation with CT scan, particularly in the first and second trimester, some centers prefer Magnetic resonance imaging as the diagnostic modality of choice. However, it is unsure at this time if it is safe to use Gadolinium. There is evidence that Gadolinium can get into the placental circulation. However, it is unclear if it has any teratogenic effects on the fetus. Hence, radiological study should be used prudently considering the risk versus benefit to the mother and fetus. No adverse effects have been noted with prenatal USG in children up to 8 years old from in utero exposure.[29] Overall, it may be quite reasonable to do an USG of the liver when pregnant women have abnormal Liver enzyme elevations.

Potential Molecular Mechanisms

Vascular Remodeling and Placentation

In pregnancy, successful placentation involves the development of a low-impedance uteroplacental circulation after trophoblast invasion and transformation of the maternal intramyometrial portion of the spiral arterioles.[30] Inadequate vascular placental invasion has been the leading hypothesis in the etiology of preeclampsia-eclampsia and HELLP syndrome. In this theory, pregnancies complicated by preeclampsia-eclampsia and HELLP syndrome have abnormal trophoblast invasion of the spiral arterioles, resulting in impaired uteroplacental perfusion. This has been called the "placental hypoxia theory".[31] This in turn results in the release of factors into the maternal circulation that may be responsible in endothelial dysfunction, vasoconstriction, hypertension and ultimately organ disease, including the liver.[32]

Vascular endothelial growth factor (VEGF) is a disulphide-linked hemodimeric glycoprotein produced by a variety of cell types, including the placenta.[33] It is selectively mitogenic for endothelial cells and appears to play a major role in the mediation of vasculogenesis, angiogenesis, in the control of the microvascular permeability and vasodilatation.[34] Hypoxia is a potent stimulus for the induction of VEGF genes expression. It has been shown that in placentas from pregnancies with hypertensive disorders, a hypoperfusion occurs followed by a hypoxic environment that stimulates the VEGF production. The biological activity of VEGF is regulated by a soluble portion of the fms-like tyrosine kinase (Flt-1) receptor (sFlt-1).[35] This protein adheres to the receptor-binding domains of placental growth factor (PIGF) and VEGF, preventing their interaction with endothelial receptors on the cell surface and thus inducing endothelial cell dysfunction.[34] Circulating levels of sFlt-1 and VEGF significantly increase as gestation progresses and are further elevated with the clinical development of preeclampsia.[33,35,36] Alterations in the levels of sFlt-1 and free VEGF were greater in women with an earlier onset of preeclampsia and predate the disease onset by several weeks.[34] Elevated concentrations of sFLT-1 in maternal plasma may be seen as early as the midtrimester.[37] This may present as an early marker for future development of preeclampsia-eclampsia and HELLP syndrome. The possible role of VEGF in predicting liver involvement and later sequelae is at present unclear.

Immunology

Preeclampsia has been suggested to be an immunologic disorder. The link between the immune system and preeclampsia-eclampsia is thought to be a maternal immune intolerance to a fetoplacental antigen.[38] Fas, also called AP0-1 or CD95, is a cell surface receptor that can induce apoptotic cell death in sensitive cells. Fas and Fas ligand-mediated apoptosis is involved in several regulatory functions within the immune system and it is proposed to be involved in the development of both preeclampsia-eclampsia and HELLP syndrome.[39] The typical hepatic lesion associated with HELLP syndrome is a periportal and/or focal parenchymal hepatocyte destruction. It has been proposed that the hepatocyte destruction involves an abnormal regulation of apoptosis and the Fas or CD95 receptor-ligand system may be involved.[39] Elevated serum soluble Fas has been seen in preeclampsia and HELLP syndrome.[40,41] Such elevations might indicate protection of maternal T-lymphocyte apoptosis and consequently lead to the maternal immune intolerance noted in these disorders. The source of elevated serum levels of soluble Fas in preeclampsia-eclampsia and HELLP syndrome remains to be determined. Further work by Strand et al have found that Fas-ligand derived from the placenta acts systemically and is a primary cause of liver damage in HELLP syndrome.[38] Blocking of the Fas-ligand can reduce liver cell apoptosis and may present a future therapeutic advance.

Fatty Acid Oxidation Defects and HELLP Syndrome

Diseases unique to pregnancy, including preeclampsia and HELLP syndrome have been shown in some cases to be associated with defects in β-oxidation of fatty acids including long chain 3-hydroxyacyl-CoA dehydrogenase,[42-51] carnitine palmitoyl transferase I,[52] medium chain acyl-CoA dehydrogenase,[53] and short chain acyl-CoA dehydrogenase,[54] deficiencies. The association between fatty acid oxidation defects and acute fatty liver of pregnancy (AFLP) is well documented in families with known pediatric LCHAD deficiency.[48,49,51] Evidence suggests that 15-20% of women with AFLP develop the maternal illness secondary to fetal LCHAD deficiency.[55] The association between fatty aid oxidation defects and HELLP syndrome is not as documented as that with AFLP. Because of the difficulty in differentiating

AFLP from HELLP in absence of liver biopsy and presence of DIC, it is possible that some cases labeled as HELLP syndrome are in fact AFLP. Yang et al prospectively screened 81 women from the US who developed HELLP syndrome and their newborns for mutations in mitochondrial trifunctional protein, and although one woman was heterozygous for the common mutation G1528C, none of the offspring had mutations in the MTP α-subunit.[55] Further, two other studies in the Finish and Italian populations reached similar conclusions. In another study,[56] den Boer et al screened 113 women with HELLP syndrome in Netherlands for the common LCHAD G1528C mutation. Only one woman was heterozygous for the LCHAD common mutation. In another study, Holub et al analyzed dried blood spots obtained from 88 infants born to women with HELLP syndrome in Austria for acylcarnitine profile using tandem mass spectrometry and for the common LCHAD mutation using restriction fragment length polymorphism.[57] There were no cases with fatty acid oxidation defects or common LCHAD mutations were detected in this study. To reconcile the results from these negative population studies in families with maternal HELLP syndrome and the case reports of women who had HELLP syndrome with fetal fatty acid oxidation defects, one must conclude that less than 1% of women who develop HELLP syndrome carry fetuses with fatty acid oxidation defects.

The mechanism of this rare association between fetal fatty acid oxidation defects and HELLP syndrome is yet to be clearly elucidated. The heterozygous mother is not symptomatic until she becomes pregnant with a fetus who is homozygous for the defect. The accumulation of potentially toxic intermediate products of fatty acid metabolism in the mother can theoretically occur from three sources: the heterozygous mother herself, the homozygous fetus, or the homozygous placenta, which has the same genetic makeup as the fetus. The mother seems an unlikely source because this would imply that HELLP syndrome should occur in metabolically stressed nonpregnant female and male heterozygotes. The homozygous fetus is unlikely to produce intermediates of fatty acid oxidation because glucose is the main energy source for the fetus, and fetal fatty acid oxidation is low.[59] It has been shown that placenta expresses the active enzymes of fatty acid oxidation.[60,61] Thus, it is possible that the placenta is the source for the toxic metabolites that cause the systemic manifestations in HELLP syndrome.

How the accumulated intermediates of fatty acid oxidation translate into maternal diseases is yet to be demonstrated. Hypothetically, these intermediates may act as free radicals causing damage to cell membranes and organelles. In LCHAD deficiency, the accumulated metabolic intermediates include long-chain 3-hydroxy-fatty acids, 3-hydroxyacylcarnitines, 3-hydroxyacyl-CoAs, and 3-hydroxy-dicarboxylic acids, which in high concentrations can injure cell membranes, potentiate free radical-induced lipid peroxidation, inhibit Na^+-K^+-ATPase, uncouple mitochondrial oxidative phosphorylation, and damage mitochondria.[62-65] Widespread damage to the maternal endothelium may cause the release of inflammatory mediators, leading to a systemic illness with multiple organ damage. In fact, damage to vascular endothelium may be an early event in the pathophysiology of preeclampsia, and oxidative stress is favored as the cause for endothelial damage in preeclampsia.[66-68] The origin of this oxidative stress may be in the placenta. It has been suggested that the placenta may be an important source of lipid peroxides in preeclampsia. Placental mitochondria may contribute to the abnormal increase in lipid peroxidation that occurs in preeclamptic placentas by an increase in mitochondrial susceptibility to lipid peroxidation.[66,69] The trophoblastic mitochondria show swelling and a loss of cristae, a change that has also been detected in mitochondria from maternal tissues in the setting of preeclampsia.[70,71] Figure 1 lists potential mechanisms of the pathogenesis of Preeclampsia/HELLP syndrome.

Pathophysiological stages of preeclampsia & HELLP syndrome

Stage 1: Abnormal placentation

Poor endometrium and inner (junctional zone) myometrium preparation resulting in impaired trophoblast invasion of myometrial artery and poor spiral artery adaptation

↓

Trigger

Ischemia-reperfusion causing placental oxidative and endoplasmic reticulum stress

↓

Stage 2: Release of mediators and associated inflammatory response

Release components like sFlt-1 and other mediators, from the intervillous space in to systemic maternal circulation resulting in intravascular systemic inflammatory response. This inflammatory response may result in generalized endothelial dysfunction, leukocyte and complement activation, decreased intravascular volume, increased vascular reactivity and multi organ dysfunction

Figure 1. Pathophysiological stages of preeclampsia and HELLP syndrome.

Management

The management of severe preeclampsia and HELLP syndrome is similar. Patients who are remote from term should be referred to a tertiary care center. The first priority is to assess and stabilize the maternal condition, particularly coagulation abnormalities. The next step is in the evaluation of fetal well-being and gestational age. Finally, a decision must be made as to whether or not immediate delivery is indicated.

Because of the association of this syndrome with high maternal morbidity and mortality some authors consider this syndrome to be an indication for immediate delivery.[72] Fetal wellbeing prior to delivery is critical to reduce the neonatal morbidity and mortality related to pre mature delivery. There is a consensus of opinion that prompt delivery is indicated if the syndrome develops after 34 weeks of gestation or earlier if there is multi-organ dysfunction, DIC, liver infarction or hemorrhage, renal failure, suspected abruption of placenta, or nonreassuring fetal status.[6,73,74] In all these cases it is considered safe to deliver considering that either the fetus is mature enough to be delivered or the pregnant woman has a serious risk of death or serious morbidity. There is significant disagreement regarding management of women with HELLP syndrome before 34 weeks of gestation. Fetal lung maturity is not achieved by this time. Some authors recommend prolonging pregnancy until 34 weeks of gestation or until the development of maternal or fetal indications for delivery.[10,14,15,24,73,75] Risk stratification or classification of HELLP syndrome may help with better understanding the characteristics of patients who may benefit from expectant management. Few studies suggest that transient improvement in laboratory values is possible with expectant management in a select group of women with this syndrome. However, most of the women in these case reports were delivered within one week of expectant management.[10,12-15,24,75] Various other treatments have been studied to lower the risk of maternal and fetal morbidity and mortality. Although it appears that expectant management may be beneficial, the overall perinatal outcome did not seem to improve when compared with cases of similar gestational age who were delivered within 48 hours after the diagnosis of HELLP syndrome.[76] Universally agreed upon recommendations would include bed rest and control of hypertension.

The lack of an effective treatment means that severe preeclampsia and HELLP syndrome require immediate delivery. In cases of extreme prematurity, when the maternal condition is stable, administration of steroids to first promote fetal lung maturity can be undertaken. However, except at the earliest gestational age, there is rarely an indication to wait to deliver as delivery of the fetus is the only known successful treatment.

The use of corticosteroids in the management of HELLP syndrome remains controversial. The beneficial effect of corticosteroids was first published in 1984, by Thiagarajah and colleagues at the University of Virginia.[17] They studied five patients with HELLP syndrome, and all five demonstrated improvements in laboratory abnormalities following high dose dexamethasone administration. Since that time, several small studies have also demonstrated an improvement in laboratory abnormalities with the use of corticosteroids. The study numbers, however, are small, and the question of whether the course of the disease is affected remains unanswered.[18,77-79] In 1999, Tompkins and Thiagarajah published the largest study to date.[16] They studied 93 patients with hematologic abnormalities associated with HELLP syndrome. All were given intramuscular injections of either betamethasone or dexamethasone. A statistically significant increase in the platelet count and a decrease in liver enzyme abnormalities were seen in all patients given corticosteroids as compared to controls. In a subsequent study by Varol et al, the postpartum course of twenty patients was studied.[79] Those receiving corticosteroids had a quicker recovery of laboratory abnormalities, decreased mean number of blood transfusions and overall shortened length of hospital stay. Although this appears promising, it remains unclear whether corticosteroids are able to alter the natural progression of liver disease associated with preeclampsia and HELLP syndrome. The use of corticosteroids to improve pregnancy outcome in women with HELLP syndrome has gained considerable interest. In a Cochran review of randomized controlled trials comparing any corticosteroid with placebo, no treatment, or other drug; or comparing one corticosteroid with another corticosteroid or dosage in women with HELLP syndrome.[75] Eleven trials (n = 550) compared corticosteroids with placebo or no treatment. There was no difference in the risk of maternal death (risk ratio (RR) 0.95, 95% confidence interval (CI) 0.28 to 3.21), maternal death or severe maternal morbidity (RR 0.27, 95% CI 0.03 to 2.12), or perinatal/infant death (RR 0.64, 95% CI 0.21 to 1.97). The only clear effect of treatment on individual outcomes was improved platelet count. The effect on platelet count was strongest for women who commenced treatment antenatally. Two trials (n = 76) compared dexamethasone with betamethasone. There was no clear evidence of a difference between groups in respect to perinatal/infant death or severe perinatal/infant morbidity or death. Maternal death and severe maternal morbidity were not reported. In respect to platelet count, dexamethasone was superior to betamethasone, both when treatment was commenced antenatally and postnatally. Authors concluded that there was no clear evidence of any effect of corticosteroids on substantive clinical outcomes. There are substantial differences in methodology and time of administration in these studies. Those receiving steroids showed significantly greater improvement in platelet counts which was greater for those receiving dexamethasone than those receiving betamethasone. The question of whether corticosteroids actually improve organ function or just alter laboratory abnormalities remains unanswered. Improvement in platelet count with corticosteroid use may make the pregnant women eligible for epidural anesthesia during delivery. The regimen of steroids used in these studies includes intravascular dexamethasone with varying dose and frequency or 2 doses of intramuscular betamethasone 12mg either 12 or 24 hours apart. The recommended regimens of corticosteroid for the enhancement of fetal maturity are 2 doses of intramuscular betamethasone 12 mg every 24 hours or 4 doses of intramuscular dexamethasone 6 mg every 12 hours.[75]

The question of whether corticosteroids actually improve liver function or just alter liver enzyme abnormalities remains unanswered.

Because HELLP syndrome is often not accompanied by diagnostic signs and symptoms like preeclampsia, it can be missed and is often advanced before an accurate diagnosis is made. Martin et al reported on the natural progression of HELLP syndrome.[24] In this study, the charts of 158 patients managed in a single tertiary center where reviewed. Inclusion criteria for HELLP syndrome was strict and included thrombocytopenia (<100,000/mm^3) with a documented normal prenatal platelet count, evidence of hepatic dysfunction as documented in elevated liver enzymes, and evidence of intravascular hemolysis. Finally, there had to be no evidence of another disorder causative of the laboratory and clinical findings.[24] They found that the disease appears to achieve peak intensity during the 24 to 48 hours after delivery. Many factors affect the time course of recovery from HELLP syndrome, however, the most important factor appears to be how quickly the disease is diagnosed and when, in the course of disease progression, the pregnancy termination occurs. Decreasing platelet counts reach a nadir at 24-48 hours after delivery. Likewise, lactate dehydrogenase concentrations and liver enzymes reach a peak at 24-48 hours postpartum. They concluded that an upward trend in platelet count and a downward trend in liver enzyme concentrations should be apparent in patients without complications by the fourth postpartum day. In all patients who recovered, a platelet count >100,000, was spontaneously achieved by the sixth postpartum day or within 72 hours of the platelet nadir.

The liver, brain, and kidney are the three main target organs damaged in severe preeclampsia and HELLP syndrome. Some of the most serious complications involve the liver, leading to life threatening liver hematomas and possible rupture of the liver capsule. In fact, 80% of women with spontaneous hepatic hemorrhage in pregnancy are toxemic.[76] Rupture of the liver is preceded by a parenchymal hematoma, almost always in the right lobe. This detaches, elevates, and eventually tears the capsule resulting in exsanguinating hemorrhage. Even the intact hematoma may prove fatal. Unfortunately, abnormalities in liver function tests are not a good predictor of abnormal hepatic imaging findings and risk of liver rupture. A study of 34 patients by Barton and Sibai found that the severity of liver enzyme abnormalities was not correlated with the abnormal hepatic imaging findings of subcapsular hematoma (n = 13) or intraparenchymal hemorrhage (n = 6).[28] These numbers, although small, reflect the difficulty in predicting the clinical outcome by the severity of liver enzyme abnormalities. The degree of thrombocytopenia was a better predictor of abnormal liver findings, particularly with platelet counts of $<20 \times 10^9$/L.

Hepatic rupture, although rare, is perhaps the most catastrophic complication of HELLP syndrome. Specific treatment options are currently limited and not routinely studied. Several surgical treatments have been described including hepatic artery ligation, hepatic packing or lobectomy, arterial embolization, and liver transplantation.[80] In a review by Smith et al at a large teaching institution and referral center, the incidence of liver rupture was one per 45,145 births.[80] Over a span of eleven years, twenty eight cases were identified. Most cases were managed by packing and drainage, achieving an overall survival rate of 82%, whereas those requiring a lobectomy had a survival rate of just 25%. The authors recommend that more aggressive surgical techniques should be reserved for refractory cases. In 1990, Terasaki described successful transcatheter embolization of the hepatic artery with gelatin particles in four patients with spontaneous hepatic rupture.[81]

Table 5. Maternal and fetal complications of preeclampsia

Maternal Complications
Eclampsia
Hemolysis, Elevated liver functions tests, and Low platelet counts (HELLP) syndrome
Hepatic infarction
Hepatic rupture
Stroke
Acute renal failure
Pulmonary edema
Abruptio placentae
Cerebral edema and herniation
Cerebral hemorrhage
Retinal detachment
Laryngeal edema
Acute respiratory distress syndrome
Fetal Complications
Preterm labor
Prematurity
Intrauterine Growth Restriction (IUGR)
Fetal death

Liver transplantation after a massive spontaneous hepatic rupture in pregnancies complicated by preeclampsia have also been performed. Hunter et al describe a liver transplantation done in desperation following a lifesaving total hepatectomy.[82] A portocaval shunt was completed to allow for venous decompression while waiting for an available orthotopic liver. Other successful liver transplantations after hepatic rupture in severe preeclampsia or HELLP syndrome have been described both in the United States and in Europe.[83,84]

A second common and devastating consequence of liver involvement in severe preeclampsia and HELLP syndrome is liver insufficiency. In fact, these diseases can often be mistaken for disseminated intravascular coagulation (DIC) as hematologic changes may develop similar to what one would expect with DIC. A similar spectrum of thrombocytopenia, prolongation of the thrombin time, hemolysis and deposition of fibrin intravascularly has been described in fatal cases.[85] Autopsies on patients who died of preeclampsia or HELLP syndrome reveal hemorrhagic necrosis of several organs due to fibrin and platelet deposition in small vessels.[86] Thus, it appears that the coagulation changes when present in preeclampsia and HELLP syndrome are similar to DIC, and demonstrate a severe form of the disease.[87,88]

Other serious maternal complications include abruptio placentae, acute renal failure, pulmonary edema, cerebral edema and herniation, cerebral hemorrhage, retinal detachment, laryngeal edema, and acute respiratory distress syndrome (Table 5). The most common cause of death is central nervous system catastrophes. All women with diagnosed preeclampsia and HELLP syndrome should receive IV infused magnesium sulfate for the prevention of eclamptic seizures. Cerebral changes must be followed closely and treated aggressively.

Future Research/Genetics

Although the spectrum of preeclampsia and HELLP syndrome are of major obstetric importance throughout the world, it remains mystifying. After more than a century of investigation, neither the cause nor possible prevention strategies have been elucidated. Family pedigree analyses have shown that genetic factors play a role in the disease, but the exact inheritance pattern is unknown. Evidence for a genetic component comes from the observation that there is a marked increase in preeclampsia among mothers, daughters, sisters, and granddaughters of women who have had preeclampsia.[89] In addition, higher concordance rates are seen among monozygotic twins compared with dizygotic twins.[90,91] Fetal genotypes, as well as environmental factors play an important role in determining susceptibility as there is still a high discordant rate among monozygotic twins developing preeclampsia during their own pregnancies. Several other factors also suggest a fetal component such as the increased rate of preeclampsia in pregnancies complicated by chromosomal abnormalities and fetal hydrops.

It has been suggested that the preeclampsia phenotype is due to a maternal-fetal gene-gene interaction either at the same locus or at separate ones. It seems unlikely that one gene can account for all of the genetic risk in all women. Rather, it is likely that several genes are acting in both the mother and her fetus, and these genes are modified by various environmental factors.[38] Genetic predisposition can be involved in any aspect of preeclampsia etiology including immune maladaption, placental ischemia, or oxidative stress.[38] Genes involved in blood pressure regulation, placentation, and vascular remodeling/injury may be involved in causing placental ischemia, whereas genes in the endothelial cell health category may lead to oxidative stress.[38] Several candidate genes have been proposed including genes under investigation in the adult hypertension literature. Finally, there is the possibility that different genetic variants are involved in predicting which women will develop liver disease and progress to liver failure and which women will have no evidence of liver disease.

Conclusion

Despite all of the extensive research surrounding this devastating disease, many unanswered questions remain. The etiology and molecular mechanisms mitigating the liver pathology remain only speculative. Leading the investigation is the search for the early markers of the disease and disease severity. The presence and severity of liver involvement is often unknown despite the many recent technological advances. Early detection of the disease and expedited delivery to prevent maternal complications and fetal compromise remain the only available treatment. Future research will likely focus on early markers with an emphasis on the genetics of preeclampsia/eclampsia and HELLP syndrome.

References

1. National High Blood Pressure Education Program Working Group Report on High Blood Pressure in Pregnancy. Am J Obstet Gynecol 1990; 163(5 Pt 1):1691-1712.
2. Report of the National High Blood Pressure Education Program Working Group on High Blood Pressure in Pregnancy. Am J Obstet Gynecol 2000; 183(1):S1-S22.
3. Sibai BM, Mercer B, Sarinoglu C. Severe pre-eclampsia in the second trimester: recurrence risk and long-term prognosis. Am J Obstet Gynecol 1991; 165(5 Pt 1):1408-1412.
4. Weinstein L. Syndrome of hemolysis, elevated liver enzymes, and low platelet count: a severe consequence of hypertension in pregnancy. Am J Obstet Gynecol 1982; 142(2):159-167.
5. Sibai BM, Taslimi MM, el-Nazer A et al. Maternal-perinatal outcome associated with the syndrome of hemolysis, elevated liver enzymes, and low platelets in severe pre-eclampsia-eclampsia. Am J Obstet Gynecol 1986; 155(3):501-509.
6. Sibai BM, Ramadan MK, Usta I et al. Maternal morbidity and mortality in 442 pregnancies with hemolysis, elevated liver enzymes, and low platelets (HELLP syndrome). Am J Obstet Gynecol 1993; 169(4):1000-1006.

7. Anselem O, Girard G, Stepanian A et al. Influence of ethnicity on the clinical and biologic expression of pre-eclampsia in the ECLAXIR study. Int J Gynaecol Obstet 2011; 115(2):153-156.
8. Sibai BM. The HELLP syndrome (hemolysis, elevated liver enzymes, and low platelets): much ado about nothing? Am J Obstet Gynecol 1990; 162(2):311-316.
9. Baxter JK, Weinstein L. HELLP syndrome: the state of the art. Obstet Gynecol Surv 2004; 59(12):838-845.
10. MacKenna J, Dover NL, Brame RG. Pre-eclampsia associated with hemolysis, elevated liver enzymes, and low platelets—an obstetric emergency? Obstet Gynecol 1983; 62(6):751-754.
11. Aarnoudse JG, Houthoff HJ, Weits J et al. A syndrome of liver damage and intravascular co-agulation in the last trimester of normotensive pregnancy. A clinical and histopathological study. Br J Obstet Gynaecol 1986; 93(2):145-155.
12. Goodlin RC, Holdt D. Impending gestosis. Obstet Gynecol 1981; 58(6):743-745.
13. Clark SL, Phelan JR, Allen SH et al. Antepartum reversal of hematologic abnormalities associated with the HELLP syndrome. A report of three cases. J Reprod Med 1986; 31(1):70-72.
14. Heyborne KD, Burke MS, Porreco RP. Prolongation of premature gestation in women with hemolysis, elevated liver enzymes and low platelets. A report of five cases. J Reprod Med 1990; 35(1):53-57.
15. Heller CS, Elliott JP. High-order multiple pregnancies complicated by HELLP syndrome. A report of four cases with corticosteroid therapy to prolong gestation. J Reprod Med 1997; 42(11):743-746.
16. Tompkins MJ, Thiagarajah S. HELLP (hemolysis, elevated liver enzymes, and low platelet count) syndrome: the benefit of corticosteroids. Am J Obstet Gynecol 1999; 181(2):304-309.
17. Thiagarajah S, Bourgeois FJ, Harbert GM Jr et al. Thrombocytopenia in pre-eclampsia: associated abnormalities and management principles. Am J Obstet Gynecol 1984; 150(1):1-7.
18. O'Brien JM, Milligan DA, Barton JR. Impact of high-dose corticosteroid therapy for patients with HELLP (hemolysis, elevated liver enzymes, and low platelet count) syndrome. Am J Obstet Gynecol 2000; 183(4):921-924.
19. Weinstein L. Pre-eclampsia/eclampsia with hemolysis, elevated liver enzymes, and thrombocytopenia. Obstet Gynecol 1985; 66(5):657-660.
20. Audibert F, Friedman SA, Frangieh AY et al. Clinical utility of strict diagnostic criteria for the HELLP (hemolysis, elevated liver enzymes, and low platelets) syndrome. Am J Obstet Gynecol 1996; 175(2):460-464.
21. Haram K, Svendsen E, Abildgaard U. The HELLP syndrome: clinical issues and management. A Review. BMC pregnancy and childbirth 2009; 9:8.
22. Martin JN Jr, Rinehart BK, May WL et al. The spectrum of severe pre-eclampsia: comparative analysis by HELLP (hemolysis, elevated liver enzyme levels, and low platelet count) syndrome classification. Am J Obstet Gynecol 1999; 180(6 Pt 1):1373-1384.
23. Sibai BM. Diagnosis, controversies, and management of the syndrome of hemolysis, elevated liver enzymes, and low platelet count. Obstet Gynecol 2004; 103(5 Pt 1):981-991.
24. Martin JN Jr, Blake PG, Perry KG Jr, et al. The natural history of HELLP syndrome: patterns of disease progression and regression. Am J Obstet Gynecol 1991; 164(6 Pt 1):1500-1509; discussion 1509-1513.
25. Rolfes DB, Ishak KG. Liver disease in toxemia of pregnancy. Am J Gastroenterol 1986; 81(12):1138-1144.
26. Arias F, Mancilla-Jimenez R. Hepatic fibrinogen deposits in pre-eclampsia. Immunofluorescent evidence. N Engl J Med 1976; 295(11):578-582.
27. Barton JR, Riely CA, Adamec TA et al. Hepatic histopathologic condition does not correlate with laboratory abnormalities in HELLP syndrome (hemolysis, elevated liver enzymes, and low platelet count). Am J Obstet Gynecol 1992; 167(6):1538-1543.
28. Barton JR, Sibai BM. Hepatic imaging in HELLP syndrome (hemolysis, elevated liver enzymes, and low platelet count). Am J Obstet Gynecol 1996; 174(6):1820-1825; discussion 1825-1827.
29. Newnham JP, Doherty DA, Kendall GE et al. Effects of repeated prenatal ultrasound examinations on childhood outcome up to 8 years of age: follow-up of a randomised controlled trial. Lancet 2004; 364(9450):2038-2044.

30. Zhou Y, McMaster M, Woo K et al. Vascular endothelial growth factor ligands and receptors that regulate human cytotrophoblast survival are dysregulated in severe pre-eclampsia and hemolysis, elevated liver enzymes, and low platelets syndrome. Am J Pathol 2002; 160(4):1405-1423.
31. Zhou Y, Damsky CH, Fisher SJ. Pre-eclampsia is associated with failure of human cytotrophoblasts to mimic a vascular adhesion phenotype. One cause of defective endovascular invasion in this syndrome? J Clin Invest 1997; 99(9):2152-2164.
32. Roberts JM, Redman CW. Pre-eclampsia: more than pregnancy-induced hypertension. Lancet 1993; 341(8858):1447-1451.
33. Levine RJ, Maynard SE, Qian C et al. Circulating angiogenic factors and the risk of pre-eclampsia. N Engl J Med 2004; 350(7):672-683.
34. Sgambati E, Marini M, Zappoli Thyrion GD et al. VEGF expression in the placenta from pregnancies complicated by hypertensive disorders. BJOG 2004; 111(6):564-570.
35. McKeeman GC, Ardill JE, Caldwell CM et al. Soluble vascular endothelial growth factor receptor-1 (sFlt-1) is increased throughout gestation in patients who have pre-eclampsia develop. Am J Obstet Gynecol 2004; 191(4):1240-1246.
36. Galazios G, Papazoglou D, Giagloglou K et al. Umbilical cord serum vascular endothelial growth factor (VEGF) levels in normal pregnancies and in pregnancies complicated by preterm delivery or pre-eclampsia. Int J Gynaecol Obstet 2004; 85(1):6-11.
37. Park CW, Park JS, Shim SS et al. An elevated maternal plasma, but not amniotic fluid, soluble fms-like tyrosine kinase-1 (sFlt-1) at the time of mid-trimester genetic amniocentesis is a risk factor for pre-eclampsia. Am J Obstet Gynecol 2005; 193(3 Pt 2):984-989.
38. Dekker GA, Robillard PY, Hulsey TC. Immune maladaptation in the etiology of pre-eclampsia: a review of corroborative epidemiologic studies. Obstet Gynecol Surv 1998; 53(6):377-382.
39. Strand S, Strand D, Seufert R et al. Placenta-derived CD95 ligand causes liver damage in hemolysis, elevated liver enzymes, and low platelet count syndrome. Gastroenterology 2004; 126(3):849-858.
40. Harirah H, Donia SE, Hsu CD. Serum soluble Fas in the syndrome of hemolysis, elevated liver enzymes, and low platelets. Obstet Gynecol 2001; 98(2):295-298.
41. Hsu CD, Harirah H, Basherra H et al. Serum soluble Fas levels in pre-eclampsia. Obstet Gynecol 2001; 97(4):530-532.
42. Schoeman MN, Batey RG, Wilcken B. Recurrent acute fatty liver of pregnancy associated with a fatty-acid oxidation defect in the offspring. Gastroenterology 1991; 100(2):544-548.
43. Wilcken B, Leung KC, Hammond J et al. Pregnancy and fetal long-chain 3-hydroxyacyl co-enzyme A dehydrogenase deficiency. Lancet 1993; 341(8842):407-408.
44. Treem WR, Rinaldo P, Hale DE et al. Acute fatty liver of pregnancy and long-chain 3-hydroxyacyl-coenzyme A dehydrogenase deficiency. Hepatology 1994; 19(2):339-345.
45. Pollitt RJ. Disorders of mitochondrial long-chain fatty acid oxidation. J Inherit Metab Dis 1995; 18(4):473-490.
46. Sims HF, Brackett JC, Powell CK et al. The molecular basis of pediatric long chain 3-hydroxyacyl-CoA dehydrogenase deficiency associated with maternal acute fatty liver of pregnancy. Proc Natl Acad Sci U S A 1995; 92(3):841-845.
47. Isaacs JD Jr, Sims HF, Powell CK et al. Maternal acute fatty liver of pregnancy associated with fetal trifunctional protein deficiency: molecular characterization of a novel maternal mutant allele. Pediatr Res 1996; 40(3):393-398.
48. Tyni T, Ekholm E, Pihko H. Pregnancy complications are frequent in long-chain 3-hydroxyacyl-coenzyme A dehydrogenase deficiency. Am J Obstet Gynecol 1998; 178(3):603-608.
49. Ibdah JA, Bennett MJ, Rinaldo P et al. A fetal fatty-acid oxidation disorder as a cause of liver disease in pregnant women. N Engl J Med 1999; 340(22):1723-1731.
50. Matern D, Hart P, Murtha AP et al. Acute fatty liver of pregnancy associated with short-chain acyl-coenzyme A dehydrogenase deficiency. J Pediatr 2001; 138(4):585-588.
51. Yang Z, Zhao Y, Bennett MJ et al. Fetal genotypes and pregnancy outcomes in 35 families with mitochondrial trifunctional protein mutations. Am J Obstet Gynecol 2002; 187(3):715-720.
52. Innes AM, Seargeant LE, Balachandra K et al. Hepatic carnitine palmitoyltransferase I deficiency presenting as maternal illness in pregnancy. Pediatr Res 2000; 47(1):43-45.

53. Nelson J, Lewis B, Walters B. The HELLP syndrome associated wiht fetal medium-chain acyl-CoA dehydrogenase deficiency. J Inherit Metab Dis 2000; 23(5):518-519.
54. Matern D, Schehata BM, Shekhawa P et al. Placental floor infarction complicating the pregnancy of a fetus with long-chain 3-hydroxyacyl-CoA dehydrogenase (LCHAD) deficiency. Mol Genet Metab 2001; 72(3):265-268.
55. Yang Z, Yamada J, Zhao Y et al. Prospective screening for pediatric mitochondrial trifunctional protein defects in pregnancies complicated by liver disease. JAMA 2002; 288(17):2163-2166.
56. den Boer ME, Ijlst L, Wijburg FA et al. Heterozygosity for the common LCHAD mutation (1528g > C) is not a major cause of HELLP syndrome and the prevalence of the mutation in the Dutch population is low. Pediatr Res 2000; 48(2):151-154.
57. Holub M, Bodamer OA, Item C et al. Lack of correlation between fatty acid oxidation disorders and haemolysis, elevated liver enzymes, low platelets (HELLP) syndrome? Acta Paediatr 2005; 94(1):48-52.
58. Ibdah JA. Acute fatty liver of pregnancy: an update on pathogenesis and clinical implications. World J Gastroenterol 2006; 12(46):7397-7404.
59. Herrera E, Amusquivar E. Lipid metabolism in the fetus and the newborn. Diabetes Metab Res Rev 2000; 16(3):202-210.
60. Rakheja D, Bennett MJ, Foster BM et al. Evidence for fatty acid oxidation in human placenta, and the relationship of fatty acid oxidation enzyme activities with gestational age. Placenta 2002; 23(5):447-450.
61. Shekhawat P, Bennett MJ, Sadovsky Y et al. Human placenta metabolizes fatty acids: implications for fetal fatty acid oxidation disorders and maternal liver diseases. Am J Physiol Endocrinol Metab 2003; 284(6):E1098-E1105.
62. Kramer JH, Weglicki WB. Inhibition of sarcolemmal Na+-K+-ATPase by palmitoyl carnitine: potentiation by propranolol. Am J Physiol 1985; 248(1 Pt 2):H75-81.
63. Mak IT, Kramer JH, Weglicki WB. Potentiation of free radical-induced lipid peroxidative injury to sarcolemmal membranes by lipid amphiphiles. J Biol Chem 1986; 261(3):1153-1157.
64. Singh AK, Yoshida Y, Garvin AJ et al. Effect of fatty acids and their derivatives on mitochondrial structures. J Exp Pathol 1989; 4(1):9-15.
65. Wojtczak L, Schonfeld P. Effect of fatty acids on energy coupling processes in mitochondria. Biochim Biophys Acta 1993; 1183(1):41-57.
66. Gratacos E. Lipid-mediated endothelial dysfunction: a common factor to pre-eclampsia and chronic vascular disease. Eur J Obstet Gynecol Reprod Biol 2000; 92(1):63-66.
67. Roberts JM, Cooper DW. Pathogenesis and genetics of pre-eclampsia. Lancet 2001; 357(9249):53-56.
68. Roberts JM, Hubel CA. Is oxidative stress the link in the two-stage model of pre-eclampsia? Lancet 1999; 354(9181):788-789.
69. Wang Y, Walsh SW. Placental mitochondria as a source of oxidative stress in pre-eclampsia. Placenta 1998; 19(8):581-586.
70. Shanklin DR, Sibai BM. Ultrastructural aspects of pre-eclampsia. II. Mitochondrial changes. Am J Obstet Gynecol 1990; 163(3):943-953.
71. Shanklin DR, Sibai BM. Ultrastructural aspects of pre-eclampsia. I. Placental bed and uterine boundary vessels. Am J Obstet Gynecol 1989; 161(3):735-741.
72. Rath W, Loos W, Kuhn W et al. The importance of early laboratory screening methods for maternal and fetal outcome in cases of HELLP syndrome. Eur J Obstet Gynecol Reprod Biol 1990; 36(1-2):43-51.
73. Sibai BM. Diagnosis, controversies, and management of the syndrome of hemolysis, elevated liver enzymes, and low platelet count. Obstet Gynecol 2004; 103(5 Pt 1):981-991.
74. Haddad B, Barton JR, Livingston JC et al. Risk factors for adverse maternal outcomes among women with HELLP (hemolysis, elevated liver enzymes, and low platelet count) syndrome. Am J Obstet Gynecol 2000; 183(2):444-448.
75. Effect of corticosteroids for fetal maturation on perinatal outcomes. NIH Consensus Development Panel on the Effect of Corticosteroids for Fetal Maturation on Perinatal Outcomes. JAMA 1995; 273(5):413-418.

76. Bis KA, Waxman B. Rupture of the liver associated with pregnancy: a review of the literature and report of 2 cases. Obstet Gynecol Surv 1976; 31(11):763-773.
77. Magann EF, Martin JN Jr. Critical care of HELLP syndrome with corticosteroids. Am J Perinatol 2000; 17(8):417-422.
78. Mecacci F, Carignani L, Cioni R et al. Time course of recovery and complications of HELLP syndrome with two different treatments: heparin or dexamethasone. Thromb Res 2001; 102(2):99-105.
79. Varol F, Aydin T, Gucer F. HELLP syndrome and postpartum corticosteroids. Int J Gynaecol Obstet 2001; 73(2):157-159.
80. Smith LG Jr, Moise KJ Jr, Dildy GA, 3rd et al. Spontaneous rupture of liver during pregnancy: current therapy. Obstet Gynecol 1991; 77(2):171-175.
81. Terasaki KK, Quinn MF, Lundell CJ et al. Spontaneous hepatic hemorrhage in pre-eclampsia: treatment with hepatic arterial embolization. Radiology 1990; 174(3 Pt 2):1039-1041.
82. Hunter SK, Martin M, Benda JA et al. Liver transplant after massive spontaneous hepatic rupture in pregnancy complicated by pre-eclampsia. Obstet Gynecol 1995; 85(5 Pt 2):819-822.
83. Amon E, Allen SR, Petrie RH et al. Acute fatty liver of pregnancy associated with pre-eclampsia: management of hepatic failure with postpartum liver transplantation. Am J Perinatol 1991; 8(4):278-279.
84. Erhard J, Lange R, Niebel W et al. Acute liver necrosis in the HELLP syndrome: successful outcome after orthotopic liver transplantation. A case report. Transpl Int 1993; 6(3):179-181.
85. Mc KD, Merrill SJ, Weiner AE et al. The pathologic anatomy of eclampsia, bilateral renal cortical necrosis, pituitary necrosis, and other acute fatal complications of pregnancy, and its possible relationship to the generalized Shwartzman phenomenon. Am J Obstet Gynecol 1953; 66(3):507-539.
86. Pritchard JA, Cunningham FG, Mason RA. Co-agulation changes in eclampsia: their frequency and pathogenesis. Am J Obstet Gynecol 1976; 124(8):855-864.
87. Kitzmiller JL, Lang JE, Yelenosky PF et al. Hematologic assays in pre-eclampsia. Am J Obstet Gynecol 1974; 118(3):362-367.
88. Pritchard JA, Weisman R Jr, Ratnoff OD et al. Intravascular hemolysis, thrombocytopenia and other hematologic abnormalities associated with severe toxemia of pregnancy. N Engl J Med 1954; 250(3):89-98.
89. Arngrimsson R, Bjornsson S, Geirsson RT et al. Genetic and familial predisposition to eclampsia and pre-eclampsia in a defined population. Br J Obstet Gynaecol 1990; 97(9):762-769.
90. Wilson ML, Goodwin TM, Pan VL et al. Molecular epidemiology of pre-eclampsia. Obstet Gynecol Surv 2003; 58(1):39-66.
91. Lachmeijer AM, Dekker GA, Pals G et al. Searching for pre-eclampsia genes: the current position. Eur J Obstet Gynecol Reprod Biol 2002; 105(2):94-113.

Chapter 7

Acute Fatty Liver of Pregnancy

Abhishek Choudhary, Jerry Angdisen, Ghassan M. Hammoud and Jamal A. Ibdah*

Abstract

Acute fatty liver of pregnancy (AFLP) is a serious maternal illness occurring in the third trimester of pregnancy with significant perinatal and maternal mortality. Till recently, it has been considered a mysterious illness. AFLP may result from mitochondrial defects in the beta-oxidation of fatty acids, in particular a deficiency of the long-chain 3-hydroxyacyl-CoA dehydrogenase (LCHAD) in the fetus. In this chapter we review the recent advances in understanding the pathogenesis of AFLP and the causative association for the development of AFLP. Furthermore, we discuss the impact of these recent advances on the offspring born to women who develop AFLP, such that screening for a genetic defect can be lifesaving to the newborn and would allow genetic counseling in subsequent pregnancies. The molecular basis and underlying mechanism for this unique fetal-maternal interaction causing maternal liver disease is also discussed.

Introduction

Acute fatty liver of pregnancy (AFLP) is a maternal liver disease unique to pregnancy. It was first described in 1872 as "yellow acute atrophy of the liver"[1] and was described as a specific clinical entity in 1900.[2] AFLP is a rare, serious condition that occurs in the third trimester. It carries a significant perinatal and maternal mortality.[3,4] AFLP can lead to hepatic failure and encephalopathy and, if diagnosis is delayed, death for the fetus and mother can ensue. Clinical findings in AFLP vary and its diagnosis is complicated by a significant overlap in clinical and biochemical features with the HELLP (hemolysis, elevated liver enzymes, and low platelets) syndrome.[5] The cause for AFLP is unknown. Recent molecular advances suggest that AFLP may result from mitochondrial dysfunction.[5] Some of the clinical and pathological features of AFLP are similar to those found in certain autosomal, recessively inherited disorders of fatty acid oxidation (FAO).[6] Other diseases characterized by fatty liver have been shown to be associated with defects in mitochondrial β-oxidation of fatty acids. Thus, it has been suggested that AFLP may result from defects in the β-oxidation of fatty acids.[3,7] The precise mechanism of AFLP and the role of other FAO defects remain unclear and it remains to be elucidated.

Clinical Features

AFLP has an incidence of 1 per 7,270 to 13,000 deliveries.[4,5,8,9] It affects women of all ages and races and there is no distinctive epidemiologic feature that has been related to geographic areas or ethnic groups.[10] The onset of AFLP is between the 30th and 38th weeks of gestation

*Corresponding Author: Jamal A. Ibdah—Department of Internal Medicine, Division of Gastroenterology and Hepatology, University of Missouri, Columbia, Missouri, USA. Email: ibdahj@health.missouri.edu

although an early occurrence at 26 weeks has been reported.[11] There are few case reports of AFLP diagnosis in the second trimester.[12-14] AFLP is more frequent in primaparous than multiparous women but can occur after multiple uneventful pregnancies.[15] Several reports have documented recurrence of AFLP in subsequent pregnancies.[16-18]

Women with AFLP usually present in the third trimester around the 35th to 36th weeks of gestation.[19] The initial manifestations are nonspecific and include headache, fatigue, nausea and vomiting.[20] Seventy percent of patients present with nausea and vomiting, whereas 50-80% complain of right upper quadrant or epigastric pain.[5,21] Jaundice usually occurs 1-2 weeks after the onset of the nonspecific symptoms[22] and is a common sign of AFLP.[10,21,23] Rarely AFLP may present as asymptomatic elevation of transaminases. Jaundice is seen on initial presentation in severe cases.[24] In a review of 35 consecutive cases between January 1996 and June 2010, the most common presenting symptom was nausea/vomiting (88.5%), jaundice (71.4%), and abdominal pain (51.4%). Maternal morbidity included hypoglycemia (94.2%), renal failure (94.2%), coagulopathy (77.1%), ascites (48.5%), and encephalopathy (40%). There were four maternal deaths (11.4%) and five perinatal deaths (12.5%).[25]

Hepatic encephalopathy occur later in the disease and should immediately alert the physician to the possibility of AFLP.[20] Abdominal tenderness is common but not specific. The liver is small and not palpable. Hypertension, edema and ascites may be present.[26] Half of the patients develop preeclampsia.[11]

Occasionally, the patient may present with signs and symptoms of eclampsia (agitation, increased thirst, premature labor, seizures). Hypertension is mild or may be absent due to decrease of peripheral vascular resistance associated with hepatic failure.

Most patients improve 1-4 weeks postpartum. The delivery maybe spontaneous, induced or a cesarean section performed. The disease rarely worsens postpartum.[27] If not terminated, the pregnancy may culminate into hepatic failure and its complications resulting in a high maternal and fetal mortality. After delivery, most patients improve slowly and a full clinical recovery may take 1-4 weeks.

AFLP can be complicated at an early stage by an upper gastrointestinal hemorrhage due to coagulation abnormalities, acute renal failure, infections, pancreatitis, or hypoglycemia.[28] In severe untreated cases, progression is rapid over a few hours or days to hepatic failure with coma, hypoglycemia, hyperammonemia, renal failure and severe coagulopathy with hemorrhage from the gastrointestinal tract or uterus leading to death of the mother and fetus.[29] Patients who develop complications may die and is often attributed to liver failure and its complications.

Diagnosis

The diagnosis of AFLP is suggested by the clinical setting (3rd trimester), history, physical examination and data from laboratory tests often confirmed by a liver biopsy.[30] Since AFLP is a medical and obstetric emergency, early recognition and prompt treatment improves both maternal and fetal survival. In most cases, reliance on clinical and laboratory findings and physician experience to make a differential diagnosis and therapeutic decisions is critical, even without a liver biopsy.[30] Although liver biopsy is a gold standard for diagnosis of AFLP, it is not routinely done due to the urgency of the clinical situation and risk of bleeding. Liver biopsy is especially helpful early in the disease or in mild cases of AFLP, when diagnosis is not clear.[31] Thus, in most cases, reliance on laboratory data without waiting for a histologically proven diagnosis is made.

Patients suspected with AFLP have their serum aminotransferase levels measured, and if elevated, a thorough laboratory evaluation is performed. A moderate elevation of the transaminases and bilirubin is usually seen in almost all patients, ranging from just slightly above normal to greater than a 1,000 U/L. The degree of liver transaminases elevation does not reflect accurately the severity of liver dysfunction. White-cell count can be elevated and peripheral blood smears may demonstrate thrombocytopenia and normoblasts.[32] Disseminated intravascular coagulopathy (DIC) is common.[28,33] Prothrombin time (PT), partial thromboplastin time (PTT) and fibrinogen maybe abnormal. Urine analysis may reveal ketones, protein or bilirubin. Blood urea nitrogen and creatinine may be elevated and uric acid increased. Serum alkaline phosphatase is 3-4 times the normal. Serum albumin is low. Ammonia level may be elevated. The blood sugar is low and hypoglycemia may occur.[20]

Histologically, AFLP is characterized by microvesicular hepatic steatosis and mitochondrial disruption is observed under electron microscopy.[3] Microscopically, hepatocytes, in the pericentral area are swollen, pale and vacuolated in appearance due to fat filled microvesicles. The centrally located nuclei are dense, widespread inflammation or necrosis is absent. Mitochondrial dense bodies are present.[20] On Oil Red O staining, one observes the cytoplasmic vesiculation as the result of microvesicular fat composed of free fatty acids.[34,35]

Electron microscopy may show microvesicular infiltration, dilated smooth endoplasmic reticulum and mitochondrial ultrastructural changes.[17,36] In severe cases lobular disarray is seen with cell dropout and fatty change may be diffuse and involve all zones. Portal inflammatory changes can be seen as well, suggestive of cholangitis; therefore histopathology should be interpreted with the consideration of clinical presentation. Imaging studies are of little value in the diagnosis of AFLP.[5] They are mainly useful for excluding other pathological processes in the liver such as hepatic ischemia, hepatic infarct, Budd-Chiari syndrome, hepatic hematoma and/or rupture. Liver ultrasound (US) and CT were inconsistent in detecting fatty infiltration of the liver in patients with AFLP; the role of MRI and MRI-spectroscopy in particular requires further investigation.

Differential Diagnosis

AFLP is often difficult to distinguish from fulminant viral hepatitis since both may present suddenly and may progress to hepatic failure.[20,37]

Moreover, clinical findings in AFLP vary and its diagnosis is complicated by a significant overlap in clinical and biochemical features with HELLP syndrome. Severe HELLP syndrome may be impossible to distinguish from AFLP. Hypertension and peripheral edema common in preeclampsia have been reported in 50% of AFLP cases.[4] Hyperbilirubinemia, hyperuricemia, coagulopathy and hypoglycemia have been reported to occur in severe preeclampsia and have been found to occur in AFLP.[3,19,38] The overlap in clinical features and laboratory findings makes it difficult to distinguish between both diseases in some cases. HELLP syndrome is a severe form of preeclampsia that threatens the pregnant mother and her fetus. HELLP syndrome occurs in 4-20% of women with severe preeclampsia.[39,40] Around 20% of patients with AFLP have features suggestive of HELLP syndrome.[3,23,41] Hypertension, peripheral edema, hyperbilirubinemia, hyperuricemia, coagulopathy, elevated liver enzymes, thrombocytopenia and hemolysis can occur in both maternal disorders. Typically, hypoglycemia, and prolongation of prothrombin time may distinguish AFLP from HELLP syndrome.[42] Histological examination of liver biopsies from HELLP syndrome cases reveal periportal hemorrhage, ischemic changes and fibrin deposition[43] while those from AFLP cases are characterized by microvesicular fatty infiltration of the liver.[5]

Due to the clinical and laboratory overlap between AFLP and HELLP syndrome in some cases, some authors have suggested that preeclampsia, HELLP syndrome and AFLP represent different stages of the same disease[9,39,43-47] and thus may share a common pathogenesis. Histopathology often can discriminate between the two disorders. Further, an immunohistological study has suggested that a cytokine and neutrophil-mediated liver injury occurs in HELLP syndrome but not in AFLP, suggesting different underlying mechanism.[48] See Table 1.

Table 1. Distinguishing features of HELLP and AFLP

Features	HELLP	AFLP
Incidence in pregnancy	0.2-0.6%	0.005-0.01%
Onset during pregnancy	Third trimester or postpartum	Third trimester or postpartum
Presence of preeclampsia	Yes	>50%
Abdominal pain	Common	Common
Clinical features	Hemolysis, thrombocytopenia, hyperuricemia	Liver failure with coagulopathy, encephalopathy, hypoglycemia, hyperuricemia and DIC
Aminotransferases	Typically >500 units/L	Typically 300-500 units/L but could be >1000u/L
Alkaline phosphatase	Exceeds those normally observed in pregnancy	3-10 fold upper limit
Bilirubin	<5 mg/dL	<5 mg/dL
LDH	Markedly elevated	Elevated
Peripheral smear	Schistocystes, spherocytes, reticulocytes	Thrombocytopenia, normoblasts
Renal impairment	Mild increase in BUN/Creatinine	Mild to moderate increase in BUN/Creatinine
Imaging	May show hepatic rupture or infarction	May show fatty infiltrate
Histology	Periportal hemorrhage, extensive necrosis and hemorrhage	Microvesicular steatosis in zone 3
Immunohistologic	Cytokine and immune-mediated	Reactive oxygen species mediated mitochondrial apoptosis
Perinatal/fetal outcomes	Perinatal mortality, prematurity	Fetal mortality 9-23%

AFLP, acute fatty liver disease of pregnancy; HELLP, hemolysis, elevated liver enzymes, and low platelet count; LDH, lactate dehydrogenase; BUN, blood urea nitrogen.

Treatment

Acute fatty liver of pregnancy typically occurs during the third trimester. It carries the risk for significant perinatal and maternal mortality. Therefore an early diagnosis and delivery, followed by close monitoring and optimized management of the impaired liver function with all associated problems are necessary to prevent maternal and fetal death. Timely delivery after initial patient stabilization is a cure, as in most pregnancy-associated liver diseases.[24,49-51] There are no reports of women recovering from AFLP prior to delivery.[52-56] The maternal and fetal mortality were reported as high as 85%[20,57] in earlier studies, but the combination of early detection, early delivery and advances in care, put estimates of current maternal and fetal mortality rates at 18% and 23%, respectively.[4,20]

Depending on the presence of complications, complete recovery from AFLP may take anywhere from days to weeks. Since clotting factors have short half-life, prothrombin time improvement is the first sign of hepatic recovery. Patients may require supportive management including dialysis, multiple transfusion, mechanical ventilation, or glucose infusion. In general, if the patient recovers from complications, there is no hepatic sequela postpartum after AFLP. Rarely, liver transplantation is needed for management of severe acute hepatic failure in the setting of AFLP.[29,58-61] Recently, plasma exchange (PE) has been attempted in clinically severe hepatopathy due to the postulate that reactive oxygen species (ROS)-mediated mitochondrial apoptosis may be its key pathogenesis.[62] Patients in the PE group had shorter durations of hepatic function recovery, intensive care unit (ICU) stay, and hospitalization than those in the case-control group, although both groups showed the same mortality.

Pathogenesis

Until recently the pathogenesis of AFLP was unknown and still has not been fully elucidated. Recent molecular advances suggest that AFLP may result from mitochondrial dysfunction. A few reports[16,23,63,64] have indicated a strong association between AFLP and a deficiency of the enzyme long-chain 3-hydroxyacyl-CoA dehydrogenase (LCHAD), a disorder of mitochondrial fatty acid β-oxidation.

Mitochondrial Fatty Acid β-Oxidation

The β-oxidation of fatty acids is the major source of energy for skeletal muscle and the heart, while the liver oxidizes fatty acids primarily under the conditions of prolonged fasting, during illness, and at periods of increased muscular activity. Fatty acid oxidation (FAO) also plays an essential role in the intermediary metabolism in the liver. The oxidation of fatty acids in the liver fuels the synthesis of ketone bodies, 3-hydroxy butyrate and acetoacetate, which are utilized as alternative sources of energy by extrahepatic organs, like the brain, when blood glucose levels are low.[65]

Mitochondrial β-oxidation of fatty acids is a complex process that consists of multiple transport steps and four enzymatic reactions resulting in the sequential removal of two-carbon, acetyl-coenzyme A units. Plasma long chain fatty acids are transported actively across the plasma membrane, esterified to coenzyme A, carried by fatty acid binding proteins through the cytoplasm to the mitochondria, and translocated across the mitochondrial inner membrane by the carnitine shuttle to the mitochondrial matrix (Fig. 1).

Once in the mitochondrial matrix, the fatty acid is sequentially cleaved, two carbons shorter, by the four reactions of the β-oxidation spiral. Each step in the spiral is catalyzed by two to four distinct enzymes, encoded by separate nuclear genes, which exhibit different overlapping substrate specificities. The first step in the spiral, as shown in Figure 2, is an acyl-CoA dehydrogenase reaction, catalyzed by very long-chain acyl-CoA dehydrogenase (VLCAD) and its homologous enzymes, long-chain acyl-CoA dehydrogenase (LCAD), medium-chain acyl-CoA dehydrogenase (MCAD) or short-chain acyl-CoA dehydrogenase (SCAD).

This dehydrogenation step forms a double bond at the 2-3 position of the substrate producing a 2,3 enoyl-CoA and generating reducing equivalents in the form of flavin adenine dinucleotide (FAD). The second step in the pathway adds a water across the double bond and is catalyzed by either a long-chain 2,3-enoyl-CoA hydratase (LCEH) or a short-chain 2,3-enoyl-CoA hydratase (SCEH) which hydrates the 2,3 enoyl-CoA across the double bond. The third step is catalyzed by a long-chain 3-hydroxyacyl-CoA dehydrogenase (LCHAD) or a short-chain 3-hydroxyacyl-CoA dehydrogenase (SCHAD) which oxidizes the 3-hydroxy position producing a 3-ketoacyl-CoA. The reaction utilizes nicotinamide dinucleotide as a cofactor. The fourth and last step in the spiral is mediated by a long-chain 3-ketoacyl-CoA thiolase (LKAT), medium-chain 3-ketoacyl-CoA

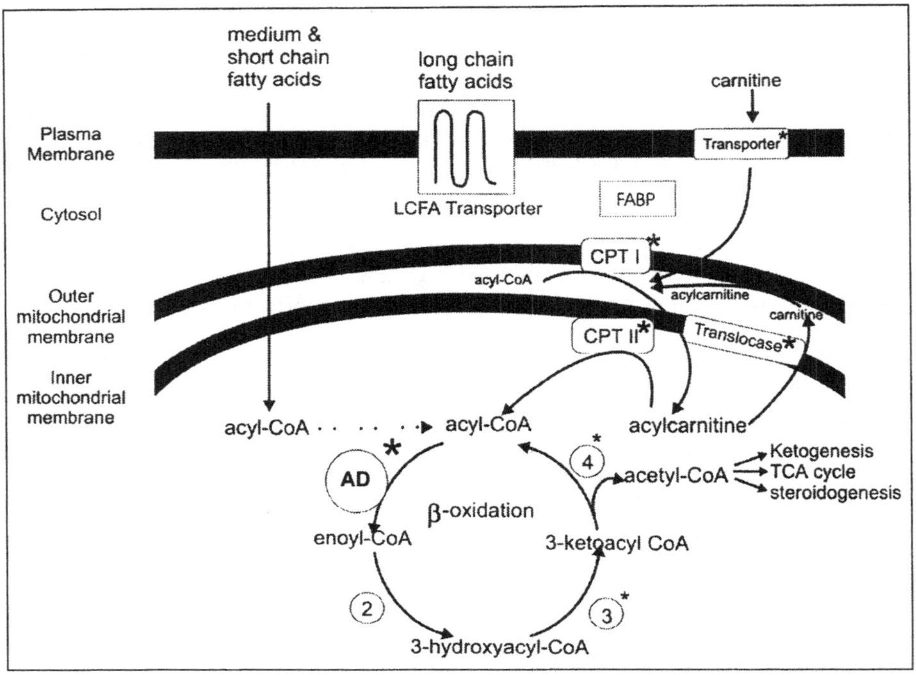

Figure 1. Fatty acid import into the mitochondria and fatty acid β-oxidation. Long chain fatty acids are actively transported across the plasma membrane, esterified to coenzyme A, carried by fatty acid binding proteins through the cytoplasm to the mitochondria and translocated across the mitochondrial membranes by the carnitine shuttle into the mitochondrial matrix. The fatty acid is subsequentially cleaved into two carbon atoms shorter by the β-oxidation spiral.

thiolase (MKAT), or short-chain 3-ketoacyl-CoA thiolase (SKAT), that shortens the fatty acyl-CoA substrate by two carbons by cleaving off acetyl-CoA. The shortened acyl-CoA can then reenter the fatty acid β-oxidation spiral until the fatty acid is completely broken down into a 2-carbon or 3-carbon specie. The acetyl-CoA produced can be used for ketogenesis, steroid genesis and as a substrate for the tricarboxylic acid cycle. The NADH and $FADH_2$ produced are used by the respiratory chain. For long chain length fatty acids, the last three steps are mediated by an enzyme complex called trifunctional protein.[66,67]

In 1992, two groups of investigators, independent of each other, reported that LCHAD is part of an enzyme complex, the mitochondrial trifunctional protein (MTP) which is associated with the inner mitochondrial membrane.[68-70] This was rapidly confirmed by other groups.[71] MTP, also known as trifunctional protein (TFP), is a heterooctamer of 4 α- and 4 β-subunits. The α-subunit amino-terminal domain contains the long-chain 3-enoyl-CoA hydratase enzymatic activity while the LCHAD enzymatic activity resides in the carboxy-terminal domain. The β-subunit has the long-chain 3-ketoacyl-CoA thiolase enzymatic activity. The association of the α- and β-subunits to form the enzyme complex is necessary for membrane translocation and for the catalytic stability of the 3 enzymes.[71,72] The human cDNAs encoding both α- and β-subunits have been isolated and characterized.[16,66] Both subunit genes, HADHA and HADHB, have been localized to chromosome 2p23 using fluorescence in situ hybridization (FISH).[73] Figure 3 shows a schematic representation of the exons of the α- and β-subunits composing the MTP subunit cDNAs and the corresponding functions of the encoded protein.

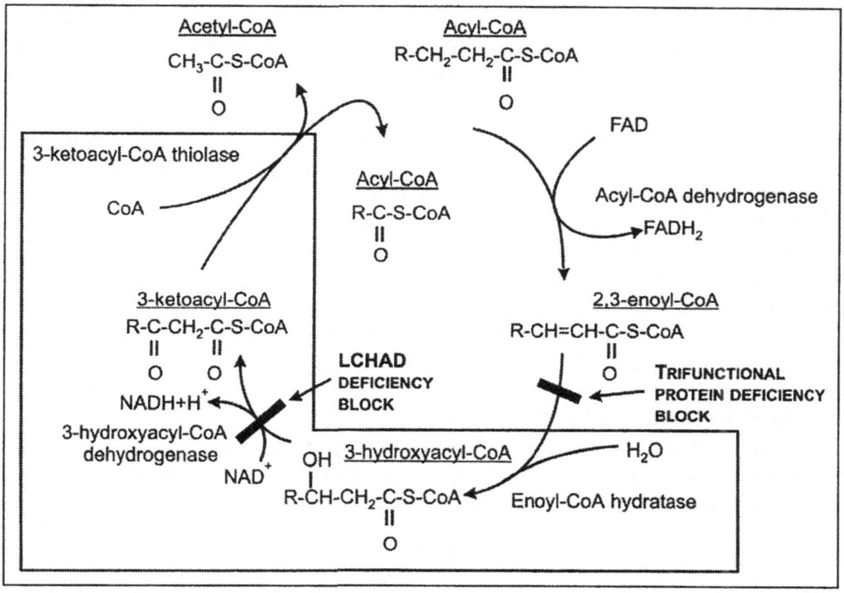

Figure 2. The biochemistry of TFP and LCHAD deficiencies. The last three reactions of the mitochondrial fatty acid β-oxidation spiral where the trifunctional protein catalyze long chain fatty acids substrates. In isolated LCHAD deficiency, the pathway is blocked after the enoyl Co-A hydratase reaction and before the 3-hydroxyacyl Co-A dehydrogenase reaction, causing the accumulation of medium- and long-chain 3-hydroxy fatty acids and their metabolites. In complete TFP deficiency, the pathway is blocked after the acyl Co-A dehydrogenase reaction and before the enoyl Co-A dehydrogenase reaction, causing the accumulation of straight chain fatty acids and their metabolites.

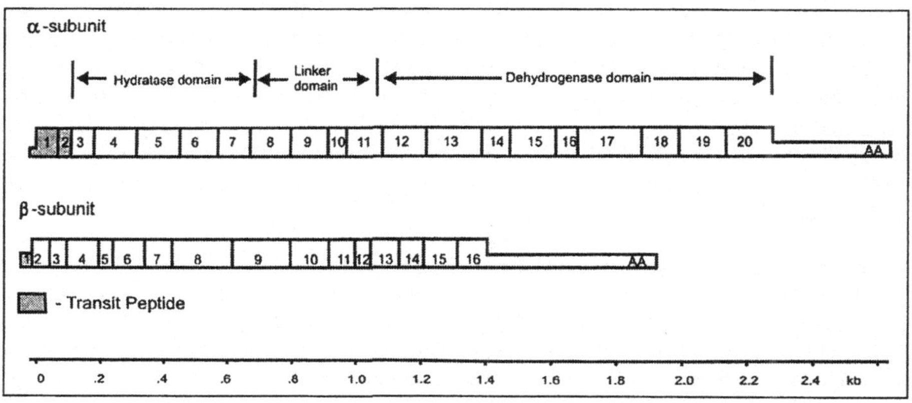

Figure 3. The structure of MTP subunit cDNA's. The MTP α-subunit contains 20 exons. Exons 1 and 2 encode the transit peptide that directs uptake into the mitochondria which is proteolytically cleaved during translocation into the mitochondria. Exons 3-8 encode the N-terminus of the mature α-subunit containing the long-chain 2,3-enoyl-CoA hydratase activity. Exon 9 encodes the region known as a linker domain. Exons 11-20 encodes the C-terminus of the mature α-subunit containing the LCHAD activity. The MTP β-subunit has 16 exons that encode the long-chain 3-keto-acyl CoA thiolase activity.

The α-subunit gene contains 20 exons spanning more than 52 kb. In the α subunit, exons 1-2 code for the transit peptide, exons 3-7 code for the hydratase domain, exons 8-10 code for the linker region and exons 11-20 code for the LCHAD domain. The β-subunit has 16 exons with exon 1 coding for the transit peptide and exons 2-16 coding for the thiolase enzymatic activity. Both genes, have been found to be arranged in a head-to-head manner on chromosome 2 as revealed by sequencing.[74] Both genes also share a promoter that is bidirectional with 2 cis elements which is activated by the transcription factor Sp1.[74]

The homologous enzymes of the MTP catalyzing shorter fatty acids are located in the mitochondrial matrix. A short-chain enoyl-CoA hydratase is responsible for the hydration of medium and short-chain enoyl-CoA species[75] a medium/short-chain 3-hydroxyacyl-CoA is responsible for medium and short-chain 3-hydroxyacl species[75] and a medium chain 3-ketoacyl thiolase activity has been described.[76] The short-chain 3-ketoacyl-CoA thiolase (SKAT) activity represents the final step in of the β-oxidation process.

Fatty Acid Oxidation Disorders

Fatty acid oxidation disorders have become an important group of inherited metabolic disorders characterized by a wide array of clinical presentations and as important causes of pediatric morbidity and mortality. The first documented inherited disorder of fatty acid oxidation was described in 1973 of a muscle carnitine palmitoyl transferase (CPT) deficiency.[77] Since then, more than 20 different disorders that affect β-oxidation have been identified. Advances in the field of fatty acid oxidation disorders have been rapid that a review article published in 1987 reported only of 7 documented inborn errors of fatty acid oxidation.[78] The most common β-oxidation defect, MCAD deficiency was first described in 1983.[79] MCAD has an approximate occurrence of 1:6,500-1:17,000 live births.[80] LCHAD deficiency was first described in 1989.[81] MTP deficiency was first reported in 1992,[82] which cleared the confusion of interpreting the combined defects of LCEH, LCHAD and LKAT.[69] VLCAD deficiency was first reported in 1993.[83,84]

Fatty acid oxidation disorders, if unrecognized and untreated, can cause sudden unexpected death. Such deaths can be certified as sudden infant death syndrome (SIDS), if fatty acid oxidation defects are not suspected. Postmortem studies on SIDS victims[84] and their siblings[85] attributed a small, but significant, number of SIDS to undiagnosed fatty acid oxidation disorders at the time of the sudden infant death. The first association of FAO disorders and SIDS was made on the basis of finding hepatic fatty infiltration in 5% of 200 SIDS cases in 1976.[86] Studies later suggested that approximately 5% of sudden infant death might be associated with fatty acid metabolic disorders.[87-89] The results of a retrospective study of 418 SIDS cases using biochemical and histochemical analyses also approximated that 5% of presumed SIDS cases may be caused by FAO disorders.[85]

Mitochondrial Trifunctional Protein Defects

MTP defects have emerged as important errors of metabolism because of their clinical implications. Human defects in the MTP complex are recessively inherited and cause either isolated LCHAD deficiency, with normal or partially reduced thiolase and hydratase activity, or complete MTP deficiency with markedly reduced activity of all 3 enzymes.[27,66,69,73,85,90-93] Patients have been described with either an isolated LCHAD deficiency or MTP deficiency.[69,85] In the literature, the majority of patients reported have been described as having isolated LCHAD deficiency.[16,91,94] Since being first described in 1989,[95] isolated LCHAD deficiency is recognized as one of the more severe FAO disorders. The estimated prevalence of MTP defects in association with fetal MTP defects in the US is 1:38,000 pregnancies.[27]

A few hours or several months after birth, children with these recessively inherited disorders present with nonketotic hypoglycemia during fasting and hepatic encephalopathy, which may progress to coma and death if untreated.[96] They can also present with cardiomyopathy, slowly progressing peripheral neuropathy, skeletal myopathy, or sudden, unexpected death.[96-100] Other phenotypes reported in the literature include hypoparathyroidism,[101] and hypocalcemia.[102,103] Late-onset neuromuscular disease and fulminant cholestatic disease have also been reported for MTP deficient patients.[98,103-105] Furthermore, higher frequencies of prematurity, intrauterine growth retardation and intrauterine death were reported in fetuses with MTP defects.[106,107]

Furthermore, our studies suggest that individuals heterozygous for MTP defects are susceptible for non-alcoholic fatty liver disease (NAFLD), the most common form of liver disease in the United States.[108]

NAFLD has a prevalence range of 15-39%.[109,110] It has a spectrum of diseases ranging from benign hepatic steatosis to non-alcoholic steatohepatitis that may progress to cirrhosis in significant number of patients.

Our work has been focused on elucidating the molecular basis for the pediatric and maternal phenotypes associated with MTP defects, and exploring potential novel therapies for these disorders.

Molecular Basis of MTP Defects

Mutations in the α-Subunit

With the explosion in human molecular genetics, we and others began to characterize normal and mutated MTP genes using molecular approaches to delineate disease-causing MTP mutations. Two teams, independent of each other, delineated the G1528C mutation in exon 15 of the α-subunit which alters amino acid 474 from glutamic acid to glutamine (E474Q) and replaces the acidic and negatively charged side-chain with a neutral, amide-containing residue.[16,94] IJlst and colleagues sequenced the cDNAs encoding the MTP α and β subunits and found the G1528C mutation in 24 of the 26 Dutch patients with LCHAD deficiency.[94] Using *S. cerevisiae* for expression studies of the wild-type and mutant protein, it was shown that the G1528C mutation was responsible for the loss of LCHAD activity.[91]

In a subsequent study, we have reported the α-subunit molecular defects and phenotypes in 24 patients with documented isolated LCHAD deficiency or complete MTP deficiency.[27] Of the 24 patients, 19 were diagnosed with isolated LCHAD deficiency and presented with the hepatic phenotype, 5 were diagnosed with complete trifunctional protein deficiency where three displayed the cardiac phenotype and the other two presented with the neuromuscular phenotype.[27] Patients with isolated LCHAD deficiency, presented predominantly with a Reye-like syndrome of liver dysfunction and carried the prevalent G1528C missense mutation on 1 or both alleles. Of the 19 subjects, 8 were homozygous for the common mutation G1528C while 11 were compound heterozygotes, where one allele carried the common mutation and the other allele had a mutation other than the common G1528C mutation.[27] The genotypes and phenotypes of these 19 patients are presented in Table 2.

The likely mechanism by which the E474Q mutation causes isolated LCHAD deficiency is that the mutation inactivates LCHAD directly within the catalytic domain, and hence, preserves the other MTP enzyme activities. The likely mechanism has been elucidated by Barycki and colleagues[111] based on the crystal and structural analysis of X-ray diffraction data from the human SCHAD which is highly homologous to the LCHAD domain. E170 in SCHAD, a residue that is analogous to LCHAD E474, is located in the NAD-binding domain within the active catalytic site. E170 is also in a position to interact with another residue, H158 which serves as a base abstracting a proton from the 3-hydroxy group of the substrate. Substitution of E170 in SCHAD with glutamine disrupts the electrostatic interaction between E170 and H158, which is essential for catalysis. E170Q mutation in SCHAD causes inactivation of the enzyme, similar to what happens to E474Q in LCHAD. The E474Q mutation causes isolated

Table 2. Genotype and phenotype in 24 families with MTP defects

Number of Families	Biochemical Phenotype	Pediatric Phenotype	Protein Expression	Pediatric Phenotype	Maternal Phenotype
19	LCHAD	Homozygous G1528C[8]	Normal	Hepatic[7]	AFLP[6]
					Mixed[1]
					Normal[1]
				Mixed[1]	HELLP[1]
		Compound heterozygous[11]	Reduced	Hepatic[9]	AFLP[7]
		G1528C/splice site[3]		Mixed[1]	HELLP[1]
		G1528C/stop codon[7]		Unknown[1]	Normal[3]
		G1528C/?*			
5	MTP	Homozygous[1] (A→7G, exon 7 missplice)	Absent	Cardiac	Normal
		Compound Heterozygous[2] (G + 1A, exon 3 missplice/A + 3G, exon 3 missplice)	Absent	Cardiac	Normal
		(C2026T, R640C/R640H exon 19)	Absent	Cardiac	Normal
		Homozygous (T845A, V246D)	Reduced	Neuromuscular	Normal
		Compound heterozygous (194A, I269N/C871T, stop residue 255)	Reduced	Neuromuscular	Normal

*Unknown mutation.
AFLP, acute fatty liver of pregnancy; HELLP, hemolysis elevated liver enzymes and low platelets syndrome.

LCHAD deficiency and blocks the β-oxidation which results in the accumulation of 3-hydroxy fatty acid metabolites.[111]

All the five patients with the complete MTP deficiency presented with cardiomyopathy or neuromyopathy and carried mutations other than the E474Q mutation (Table 2). Three of the five patients with MTP deficiency presented with cardiomyopathy. All three patients had splice site and missense mutations that caused the absence of the MTP protein complex.[27] This condition blocks the second step of the β-oxidation spiral that results in the accumulation of long-chain 3-enoyl fatty acid metabolites. The other two patients with complete MTP deficiency presented with neuromyopathy.[27] These two patients carried mutations in exon 9 that allowed stable MTP expression.[90] The exon 9 α-subunit mutations and the neuromuscular phenotype suggest a novel genotype-phenotype correlation. Exon 9 in the MTP α-subunit encodes a region described as a linker region that may be important in subunit interaction and octamer complex formation as suggested by crystallographic data from short-chain enoyl-CoA hydratase that showed a spiral structure of the active site and substrate binding domains that are highly conserved with the MTP N-terminus.[112] The exon 9 mutations in the α-subunit seem to result in defective subunit interactions and octamer formation.

MTP β-Subunit Defects

A published article[113] also reports heterogeneity in MTP β-subunit mutations. Three clinical phenotypes are also apparent in MTP β-subunit defects; hepatic, cardiac and neuromuscular. Of the 15 patients reported, 4 presented with the cardiac phenotype that was considered lethal, 2 with the hepatic phenotype and 9 with the mild neuromuscular phenotype. Mutational analysis revealed 16 different mutations with missense mutations occurring 12 times, 3 deletions and 1 frame shift mutation. Within days after birth, patients who presented with the cardiac phenotype died due to severe dilated cardiomyopathy. The milder mutation determined the clinical phenotype in compound heterozygotes with a null mutation in one allele and a missense mutation in the other allele.[113] Other groups have also reported mutations in few patients. One study reported a novel homozygous missense mutation, G976C, within a highly conserved region in the β-subunit of the MTP.[114] Another group detected 2 mutations in the HADHB (β-subunit) gene in 2 patients. One patient was a compound heterozygote; one allele had a nucleotide change and the other allele had an insertion that caused a deletion. The second patient was homozygous for a G1331A transition.[115]

MTP Fetal Defects and Maternal Liver Disease

Table 3 lists reported studies that document the association between fetal fatty acid oxidation disorders and maternal liver disease. These studies document strong causative association between fetal MTP defects and AFLP.

Shoeman and colleagues were first to report an association between recurrent maternal acute fatty liver of pregnancy with a fetal fatty acid oxidation disorder in two siblings who both died at 6 months of age.[17] The authors speculated that due to the finding of similar hepatic pathology of microvesicular steatosis, AFLP and fatty acid oxidation defects might share a common pathogenesis.

Other case reports in the early 90's have also associated affected infants with LCHAD deficiency to the occurrence of severe preeclampsia, HELLP syndrome or AFLP in the infant's mother during pregnancy.[18,63] Wilcken et al[18] reported 11 pregnancies in 5 mothers where 6 babies had confirmed LCHAD deficiency by enzymatic analysis of cultured skin fibroblasts and that the mothers had either AFLP or HELLP syndrome in all six pregnancies with the LCHAD-deficient fetuses. Treem et al[63] reported an LCHAD deficient child born to a pregnancy complicated by AFLP.

Table 3. Studies linking AFLP to fatty acid oxidation defects

Year	Report	Families/ Pregnancies	Maternal Diagnosis	Fetal Diagnosis: Biochemical	Fetal Diagnosis: Molecular Defect
1991	Schoeman et al. Recurrent acute fatty liver of pregnancy. Gastroenterology 1991; 100(2):544-548.	1/2	AFLP[2]	FAO disorder	Not Determined
1993	Wilcken B et al. Pregnancy and fetal long-chain 3-hydroxyacyl coenzyme A dehydrogenase deficiency. Lancet 1993; 341(8842):407-408.	5/11	AFLP[3] HELLP[2] Hyperemesis[2] Preeclampsia[1] Normal[3]	LCHAD Deficient[2] Not Determined[1] LCHAD Deficient[2] LCHAD Deficient[2] Normal[1] Normal[3]	Not Determined
1994	Treem WR et al. Acute fatty liver of pregnancy and long-chain 3-hydroxyacyl-coenzyme A dehydrogenase deficiency. Hepatology 1994; 19(2):339-345.	1/1	AFLP	LCHAD Deficient	Not Determined
1995	Sims HF et al. The molecular basis of pediatric long chain 3-hydroxyacyl-CoA dehydrogenase deficiency associated with maternal acute fatty liver of pregnancy. Proc Natl Acad Sci U S A. 1995; 92(3):841-845.	3/3	HELLP[2] Preeclampsia and HELLP[1]	LCHAD Deficient[2] LCHAD Deficient[1]	HO (G1528C)[2] Compound HT (G1528C/C1132T)[1]

continued on next page

Table 3. Continued

Year	Report	Families/ Pregnancies	Maternal Diagnosis	Fetal Diagnosis	
				Biochemical	Molecular Defect
1996	Issacs JD et al. Maternal acute fatty liver of pregnancy associated with fetal trifunctional protein deficiency: molecular characterization of a novel maternal mutant allele. Pediatr Res 1996; 40(3):393-398.	1/1	AFLP	Not Determined	Compound HT (G1528C/C1678T)
1996	Treem WR et al. Acute fatty liver of pregnancy, hemolysis, elevated liver enzymes, and low platelets syndrome, and long chain 3-hydroxyacyl-coenzyme A dehydrogenase deficiency. Am J Gastroenterol 1996; 91(11):2293-2300.	12/27	AFLP[13]	LCHAD Deficient[4] Partial Deficiency[2] Normal[1] Not Determined[6]	Not Determined
			Normal[14]	Partial Deficiency[5] Normal[1] Not Determined[8]	Not Determined
1998	Tyni T et al. Pregnancy complications are frequent in long-chain 3-hydroxyacyl-coenzyme A dehydrogenase deficiency. Am J Obstet Gynecol 1998; 178(3):603-608.	18/63	AFLP[1] HELLP[3] Preeclampsia[5] Other[a,54]	LCHAD Deficient[1] LCHAD Deficient[3] LCHAD Deficient[5] LCHAD Deficient[20] Normal[34]	Not Determined

continued on next page

Table 3. Continued

Year	Report	Families/ Pregnancies	Maternal Diagnosis	Fetal Diagnosis	
				Biochemical	Molecular Defect
1999	Ibdah JA et al. A fetal fatty-acid oxidation disorder as a cause of liver disease in pregnant women. N Engl J Med 1999; 340(22):1723-1731.	24/24	AFLP^{12}	LCHAD Deficient^{12}	HO (G1528C)^{6} Compound HT^{6} (G1528C/other^{b})
			HELLP^{3}	LCHAD Deficient^{3}	HO (G1528C)^{2} Compound HT^{1} (G1528C/other)
			Normal^{9}	LCHAD Deficient^{4}	HO (G1528C)^{1} Compound HT^{3} (G1528C/other)
				MTP Deficient^{5}	(other/other)^{5}
2000	Innes AM et al. Hepatic carnitine palmitoyltransferase I deficiency presenting as maternal illness in pregnancy. Pediatr Res 2000; 47(1):43-45.	1/2	AFLP^{1}	CPT-1 Deficient^{1}	Not Determined
			Hyperemesis Gravidarum^{1}	CPT-1 Deficiency	Not Determined
2001	Matern D et al. Acute fatty liver of pregnancy associated with short-chain acyl-coenzyme A dehydrogenase deficiency. J Pediatr 2001; 138(4):585-588.	1/1	AFLP	SCAD Deficient	HO (G625A)
2002	Maitra A et al. Absence of the G1528C (E474Q) mutation in the alpha-subunit of the mitochondrial trifunctional protein in women with acute fatty liver of pregnancy. Pediatr Res 2002; 51(5):658-661.	10/10	AFLP	Not Determined	Not Determined

continued on next page

Table 3. Continued

Year	Report	Families/ Pregnancies	Maternal Diagnosis	Fetal Diagnosis: Biochemical	Fetal Diagnosis: Molecular Defect
2002	Yang Z et al. Fetal genotypes and pregnancy outcomes in 35 families with mitochondrial trifunctional protein mutations. Am J Obstet Gynecol 2002; 187(3):715-720.	35/83	AFLP[20]	LCHAD Deficient[17]	HO (G1528C)[9]
				Not Determined[3]	Compound HT[8]
			HELLP[3]	LCHAD Deficient[3]	HO (G1528C)[2]
					Compound HT[1] (G1528C/other)
			Preeclampsia[1]	LCHAD Deficient[1]	Compound HT[1] (G1528C/other)
			Other[11]	LCHAD Deficient[3]	HO (G1528C)[3]
				Not Determined[8]	
			Normal[48]	LCHAD Deficient[8]	HO (G1528C)[3]
					Compound HT[5] (G1528C/other)
				MTP Deficient[5]	(other/other)[5]
				Normal[35]	
2002	Yang Z et al. Prospective screening for pediatric mitochondrial trifunctional protein defects in pregnancies complicated by liver disease. JAMA 2002; 288(17):2163-2166.	108/108	AFLP[27]	LCHAD Deficient[5]	HO (G1528C)[3]
					Compound-HT[2] (G1528C/other)
				Normal[22]	
			HELLP[81]	Normal[81]	

[a]Other—pregnancy induced hypertension,[3] edema,[4] jaundice,[3] intehepatic cholestasis,[3] hyperemesis,[3] prematurity,[12] small for gestational age,[14] asphyxia,[10] fetal death;[2] [b]Other—non-G1528C mutation.

AFLP, acute fatty liver of pregnancy; HELLP, hemolysis, elevated liver enzymes and low platelets syndrome; HO, homozygous; HT, heterozygous; CPT-1, carnitine palmitoyl transferase-1; SCAD, Short-chain acyl-CoA dehydrogenase.

In a subsequent report published in 1995, Sims et al delineated the molecular basis of pediatric LCHAD deficiency and its association with AFLP[16] in 3 families with LCHAD-deficient infants. The mothers suffered maternal HELLP syndrome or AFLP during pregnancies with the affected children. They used single-strand conformation variance (SSCV) analysis of the α-subunit of MTP to determine the molecular basis of LCHAD deficiency. The analysis in 2 affected children revealed the G1528C mutation on both alleles and the third child was a compound heterozygote with one allele carrying the G1528C.

Another report documented an association between pediatric LCHAD deficiency and occurrence of AFLP in the mother, Isaacs et al[64] reported a compound heterozygosity for MTP mutations in an affected child born to a pregnancy complicated by AFLP, with a novel mutation in exon 16 on one allele and the common G1528C mutation on the other allele.

In a study where we examined the association between MTP defects in children and liver disease in their mothers during pregnancy,[27] 15 of 24 women (62%) were diagnosed to have had maternal liver disease, while 9 of the 24 women had normal pregnancies, as summarized in Table 2. In 5 of the normal pregnancies, the affected infant did not have the G1528C mutation, but rather other MTP mutations. The remaining four normal pregnancies were associated with fetal LCHAD deficiency. Thus 15/19 pregnancies associated with fetal LCHAD deficiency were complicated by maternal liver disease and none of the pregnancies associated with complete MTP deficiency were associated with AFLP or HELLP syndrome. The results in this study show that when carrying a fetus that is LCHAD deficient, the mother has a 79% chance of a pregnancy complicated by AFLP or HELLP syndrome. It also provides strong evidence that a woman whose affected fetus has the G1528C on one or both alleles of the α subunit of the mitochondrial trifunctional protein is likely to have pregnancy complications with AFLP.

In a subsequent study, we evaluated fetal genotypes and pregnancy outcomes in 83 pregnancies in 35 families with documented pediatric MTP defects, 24 pregnancies were complicated by AFLP, HELLP syndrome or severe preeclampsia.[106] Of the 24 pregnancies, 20 were complicated by AFLP, 2 with HELLP syndrome and 2 with preeclampsia and in all the LCHAD-deficient fetus carried the G1528C mutation on one or both alleles. Five pregnancies had fetuses with complete MTP deficiency (none of the mutations were G1528C) but that there were no associated maternal complications in those pregnancies. This study also provides data that carrying a fetus with LCHAD deficiency is associated with a high risk for developing AFLP and other maternal complications during pregnancy.

With the growing evidence suggesting that carrying an LCHAD-deficient fetus is associated with AFLP, we have recommended that neonates born to pregnancies complicated by AFLP be tested for the common G1528C mutation and that this testing when done early after birth can be lifesaving as it may identify LCHAD-deficient children before they manifest the disease allowing early dietary intervention by institution of a diet low in fat, high in carbohydrate, and by substitution of the long chain fatty acids with medium chain fatty acids.

To further assess the significance of the association between maternal AFLP and fetal LCHAD deficiency, we prospectively screened for MTP mutations mothers who developed AFLP or HELLP syndrome and their newborn infants.[106] The molecular screening was based solely on the maternal history. We prospectively screened 27 pregnancies complicated by AFLP and 81 pregnancies complicated by HELLP syndrome. Out of the 27 women that developed AFLP, 5 carried fetuses with MTP mutations. Three were homozygous for the G1528C mutation and two were compound heterozygotes with one mutant allele carrying the common G1528C and the other mutant allele carrying a novel mutation. Only 1 woman diagnosed with HELLP syndrome was heterozygous for the G1528C mutation which was not detected in her infant. None of the children born to the 81 women diagnosed with HELLP syndrome carried MTP mutations. Our study document that in approximately one of five pregnancies complicated by AFLP, the fetus is LCHAD deficient. This strong association between AFLP and the common

G1528C mutation in the fetus is significant and hence screening the offspring of women who develop AFLP at birth for this mutation can be lifesaving.

In addition, identification of MTP mutations in the offspring of pregnancies complicated by AFLP allows genetic counseling for the mothers. Prenatal diagnosis can be performed in subsequent pregnancies to identify pregnancies at risk for development of AFLP. We performed molecular prenatal diagnosis in 11 pregnancies using chorionic villous samples and successfully identified the fetal genotype in these pregnancies confirmed by biochemical and molecular testing of the new born or aborted fetuses.[116]

Possible Hypothesis of AFLP Pathogenesis

The precise mechanism by which an LCHAD deficient fetus causes AFLP in a heterozygote mother is still unclear. However, several factors appear to contribute to this fetal-maternal interaction as illustrated in Figure 4.

First, the heterozygosity of the mother for an MTP defect reduces her capacity to oxidize long-chain fatty acids. Second, the stressful nature of pregnancy with its accompanying changes in metabolism, the increased lipolysis, and the decreased β-oxidation. We speculate that in the presence of the G1528C mutation, potentially hepatotoxic long-chain 3-hydroxyacyl fatty acid metabolites, produced by the fetus or placenta, accumulate in the maternal circulation. There is evidence for fatty acid oxidation in a normal human placenta including LCHAD and SCHAD activity.[57] Another study reported significant expression of fatty acid β-oxidation enzymes in human placenta as assessed by immunohistochemical and immunoblot analyses.[117] A recent study also showed high activity of fatty acid oxidation enzymes in human term placenta and chorionic villus samples.[118] The role of other FAO defects in the development of maternal liver disease is not clear. Maternal liver disease consistent with AFLP was associated with carnitine palmitoyl transferase I (CPT-1) deficiency have been reported. In Innes et al,[119] the patient developed

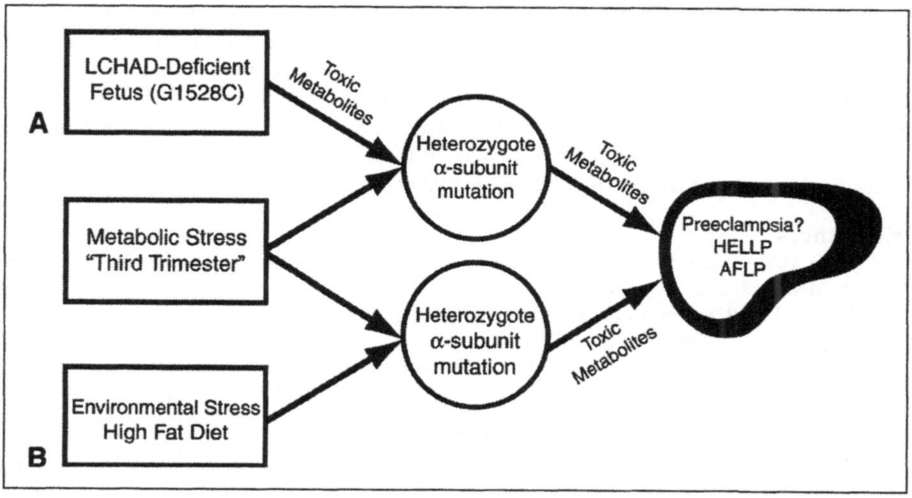

Figure 4. Hypothesis illustrating the possible role of fetal and maternal MTP mutations in developing AFLP. Carrying an LCHAD deficient fetus (A) is the major determining factor in the development of maternal illness. Hepatotoxic metabolites produced by the fetus and/or placenta may cause liver disease in the obligate heterozygous mother especially when combined with the metabolic stress of the third trimester. Environmental stress (B) may lead to the further accumulation of toxic metabolites in the genetically susceptible mother causing maternal liver disease.

liver disease consistent with AFLP and in a successive pregnancy, hyperemesis gravidarum. Both children were subsequently shown to have CPT-1 deficiency. In Ytilano et al,[120] the patient experienced complications that included a HELLP-like syndrome. She was later diagnosed to be CPT-1 deficient but her child was unaffected. Another report associated AFLP in an affected fetus with SCAD deficiency. In Matern et al,[121] an infant was evaluated for a fatty acid oxidation defect since the mother developed AFLP in the course of gestation. The infant was diagnosed as SCAD deficient. A report of a pregnancy complicated with HELLP syndrome associated with MCAD has been published. In Nelson et al,[122] the pregnancy was complicated by severe preeclampsia and the HELLP syndrome in which the fetus was subsequently shown to have MCAD deficiency. Undiagnosed maternal medium-chain acyl-CoA dehydrogenase (MCAD) deficiency in AFLP, who delivered a baby without enzyme abnormalities was also reported.[123] There is also a report of 3 patients that developed maternal liver disease that carried affected fetuses diagnosed to have MTP deficiency and carrying mutations other than the G1528C.[124] One study that looked at the incidence of maternal liver disease in pregnancies with fetal FAO defects suggested that fetal long-chain defects are 50 times more likely to develop a liver disease in pregnancy when compared with healthy controls; and short- and medium-chain FAO defects are 12 times more likely to develop liver disease.[125] These studies demonstrate a growing diversity of clinical presentations of fatty acid oxidation defects and their role in the development of maternal liver disease.

Conclusion

AFLP is an uncommon but clinically severe maternal liver disorder unique to pregnancy occurring in the third trimester with significant perinatal and maternal mortality. Mitochondrial defects in the beta-oxidation of fatty acids, in particular a deficiency of the long-chain 3-hydroxyacyl-CoA dehydrogenase in the fetus contributes to the etiology of the disease. Reactive oxygen species (ROS)-mediated mitochondrial apoptosis and oxidative stress in placental mitochondria and peroxisomes may be its key pathogenesis. The clinical findings vary and its diagnosis may be complicated by a significant overlap in clinical and biochemical features with HELLP syndrome. Early referral to a tertiary care hospital for close monitoring and optimized management of the impaired liver function with all associated problems are crucial to prevent maternal and fetal death. Timely delivery after initial patient stabilization improves the prognoses of both mother and newborn. Prenatal diagnosis is warranted in subsequent pregnancies to identify future pregnancies at risk of development of AFLP.

References

1. Jones HM. A case presenting the symptoms of acute yellow atrophy of the liver: Recovery. Br Med J 1872; 1(592):468-469.
2. Findlay JW. Remarks on the pathology of acute yellow atrophy of the liver. Br Med J 1900; 1(2057):1330-1334.
3. Riely CA. Acute fatty liver of pregnancy. Semin Liver Dis 1987; 7(1):47-54.
4. Knox TA, Olans LB. Liver disease in pregnancy. N Engl J Med 1996; 335(8):569-576.
5. Ibdah JA. Acute fatty liver of pregnancy: an update on pathogenesis and clinical implications. World J Gastroenterol 2006; 12(46):7397-7404.
6. Treem WR, Witzleben CA, Piccoli DA et al. Medium-chain and long-chain acyl CoA dehydrogenase deficiency: clinical, pathologic and ultrastructural differentiation from Reye's syndrome. Hepatology 1986; 6(6):1270-1278.
7. Sherlock S. Acute fatty liver of pregnancy and the microvesicular fat diseases. Gut 1983; 24(4):265-269.
8. Castro MA, Goodwin TM, Shaw KJ et al. Disseminated intravascular coagulation and antithrombin III depression in acute fatty liver of pregnancy. Am J Obstet Gynecol 1996; 174(1 Pt 1): 211-216.

9. Pockros PJ, Peters RL, Reynolds TB. Idiopathic fatty liver of pregnancy: findings in ten cases. Medicine (Baltimore) 1984; 63(1):1-11.
10. Treem WR. Mitochondrial fatty acid oxidation and acute fatty liver of pregnancy. Semin Gastrointest Dis 2002; 13(1):55-66.
11. Buytaert IM, Elewaut GP, Van Kets HE. Early occurrence of acute fatty liver in pregnancy. Am J Gastroenterol 1996; 91(3):603-604.
12. Monga M, Katz AR. Acute fatty liver in the second trimester of pregnancy. Primary Care Update for Ob/Gyns 1998; 5(4):191.
13. Monga M, Katz AR. Acute fatty liver in the second trimester. Obstet Gynecol 1999; 93(5 Pt 2): 811-813.
14. Suzuki S, Watanabe S, Araki T. Acute fatty liver of pregnancy at 23 weeks of gestation. BJOG 2001; 108(2):223-224.
15. Mabie WC. Obstetric management of gastroenterologic complications of pregnancy. Gastroenterol Clin North Am 1992; 21(4):923-935.
16. Sims HF, Brackett JC, Powell CK et al. The molecular basis of pediatric long chain 3-hydroxyacyl-CoA dehydrogenase deficiency associated with maternal acute fatty liver of pregnancy. Proc Natl Acad Sci U S A 1995; 92(3):841-845.
17. Schoeman MN, Batey RG, Wilcken B. Recurrent acute fatty liver of pregnancy associated with a fatty-acid oxidation defect in the offspring. Gastroenterology 1991; 100(2):544-548.
18. Wilcken B, Leung KC, Hammond J et al. Pregnancy and fetal long-chain 3-hydroxyacyl coenzyme A dehydrogenase deficiency. Lancet 1993; 341(8842):407-408.
19. Watson WJ, Seeds JW. Acute fatty liver of pregnancy. Obstet Gynecol Surv 1990; 45(9):585-591.
20. Kaplan MM. Acute fatty liver of pregnancy. N Engl J Med 1985; 313(6):367-370.
21. Vigil-De Gracia P, Lavergne JA. Acute fatty liver of pregnancy. Int J Gynaecol Obstet 2001; 72(2):193-195.
22. Matin A, Sass DA. Liver disease in pregnancy. Gastroenterol Clin North Am 2011; 40(2): 335-353, viii.
23. Treem WR, Shoup ME, Hale DE et al. Acute fatty liver of pregnancy, hemolysis, elevated liver enzymes, and low platelets syndrome, and long chain 3-hydroxyacyl-coenzyme A dehydrogenase deficiency. Am J Gastroenterol 1996; 91(11):2293-2300.
24. Mjahed K, Charra B, Hamoudi D et al. Acute fatty liver of pregnancy. Arch Gynecol Obstet 2006; 274(6):349-353.
25. Vigil-de Gracia P, Montufar-Rueda C. Acute fatty liver of pregnancy: diagnosis, treatment, and outcome based on 35 consecutive cases. J Matern Fetal Neonatal Med 2011; 24(9):1143-1146.
26. Moise KJ Jr, Shah DM. Acute fatty liver of pregnancy: etiology of fetal distress and fetal wastage. Obstet Gynecol 1987; 69(3 Pt 2):482-485.
27. Ibdah JA, Bennett MJ, Rinaldo P et al. A fetal fatty-acid oxidation disorder as a cause of liver disease in pregnant women. N Engl J Med 1999; 340(22):1723-1731.
28. Holzbach RT. Acute fatty liver of pregnancy with disseminated intravascular coagulation. Obstet Gynecol 1974; 43(5):740-744.
29. Ockner SA, Brunt EM, Cohn SM et al. Fulminant hepatic failure caused by acute fatty liver of pregnancy treated by orthotopic liver transplantation. Hepatology 1990; 11(1):59-64.
30. Wei Q, Zhang L, Liu X. Clinical diagnosis and treatment of acute fatty liver of pregnancy: a literature review and 11 new cases. J Obstet Gynaecol Res 2010; 36(4):751-756.
31. Minakami H, Takahashi T, Tamada T. Should routine liver biopsy be done for the definite diagnosis of acute fatty liver of pregnancy? Am J Obstet Gynecol 1991; 164(6 Pt 1):1690-1691.
32. Burroughs AK, Seong NH, Dojcinov DM et al. Idiopathic acute fatty liver of pregnancy in 12 patients. Q J Med 1982; 51(204):481-497.
33. Cano RI, Delman MR, Pitchumoni CS et al. Acute fatty liver of pregnancy. Complication by disseminated intravascular coagulation. JAMA 1975; 231(2):159-161.
34. Eisele JW, Barker EA, Smuckler EA. Lipid content in the liver of fatty metamorphosis of pregnancy. Am J Pathol 1975; 81(3):545-560.
35. Rolfes DB, Ishak KG. Acute fatty liver of pregnancy: a clinicopathologic study of 35 cases. Hepatology 1985; 5(6):1149-1158.

36. Barton JR, Sibai BM, Mabie WC et al. Recurrent acute fatty liver of pregnancy. Am J Obstet Gynecol 1990; 163(2):534-538.
37. Brown MS, Reddy KR, Hensley GT et al. The initial presentation of fatty liver of pregnancy mimicking acute viral hepatitis. Am J Gastroenterol 1987; 82(6):554-557.
38. Usta IM, Barton JR, Amon EA et al. Acute fatty liver of pregnancy: an experience in the diagnosis and management of fourteen cases. Am J Obstet Gynecol 1994; 171(5):1342-1347.
39. Sibai BM, Ramadan MK, Usta I et al. Maternal morbidity and mortality in 442 pregnancies with hemolysis, elevated liver enzymes, and low platelets (HELLP syndrome). Am J Obstet Gynecol 1993; 169(4):1000-1006.
40. Martin JN Jr, Blake PG, Perry KG Jr et al. The natural history of HELLP syndrome: patterns of disease progression and regression. Am J Obstet Gynecol 1991; 164(6 Pt 1):1500-1509; discussion 1509-1513.
41. Strauss AW, Bennett MJ, Rinaldo P et al. Inherited long-chain 3-hydroxyacyl-CoA dehydrogenase deficiency and a fetal-maternal interaction cause maternal liver disease and other pregnancy complications. Semin Perinatol 1999; 23(2):100-112.
42. Rahman TM, Wendon J. Severe hepatic dysfunction in pregnancy. QJM 2002; 95(6):343-357.
43. Riely CA. Hepatic disease in pregnancy. Am J Med 1994; 96(1A):18S-22S.
44. Minakami H, Oka N, Sato T et al. Preeclampsia: a microvesicular fat disease of the liver? Am J Obstet Gynecol 1988; 159(5):1043-1047.
45. Wolf JL. Liver disease in pregnancy. Med Clin North Am 1996; 80(5):1167-1187.
46. Pereira SP, O'Donohue J, Wendon J et al. Maternal and perinatal outcome in severe pregnancy-related liver disease. Hepatology 1997; 26(5):1258-1262.
47. Dani R, Mendes GS, Medeiros Jde L et al. Study of the liver changes occurring in preeclampsia and their possible pathogenetic connection with acute fatty liver of pregnancy. Am J Gastroenterol 1996; 91(2):292-294.
48. Halim A, Kanayama N, El Maradny E et al. Immunohistological study in cases of HELLP syndrome (hemolysis, elevated liver enzymes and low platelets) and acute fatty liver of pregnancy. Gynecol Obstet Invest 1996; 41(2):106-112.
49. Mjahed K, Charra B, Hamoudi D et al. Acute fatty liver of pregnancy. Arch Gynecol Obstet 2006; 274(6):349-353.
50. Ko H, Yoshida EM. Acute fatty liver of pregnancy. Can J Gastroenterol 2006; 20(1):25-30.
51. Moore M, Nelson-Piercy C. Pregnancy and the liver. Br J Hosp Med 2011; 72(11):M170-173.
52. Snyder RR, Hankins GD. Etiology and management of acute fatty liver of pregnancy. Clin Perinatol 1986; 13(4):813-825.
53. Davies MH, Wilkinson SP, Hanid MA et al. Acute liver disease with encephalopathy and renal failure in late pregnancy and the early puerperium—a study of fourteen patients. Br J Obstet Gynaecol 1980; 87(11):1005-1014.
54. Ebert EC, Sun EA, Wright SH et al. Does early diagnosis and delivery in acute fatty liver of pregnancy lead to improvement in maternal and infant survival? Dig Dis Sci 1984; 29(5):453-455.
55. Hou SH, Levin S, Ahola S et al. Acute fatty liver of pregnancy. Survival with early cesarean section. Dig Dis Sci 1984; 29(5):449-452.
56. Reyes H, Sandoval L, Wainstein A et al. Acute fatty liver of pregnancy: a clinical study of 12 episodes in 11 patients. Gut 1994; 35(1):101-106.
57. Rakheja D, Bennett MJ, Foster BM et al. Evidence for fatty acid oxidation in human placenta, and the relationship of fatty acid oxidation enzyme activities with gestational age. Placenta 2002; 23(5):447-450.
58. Luzar B, Ferlan-Marolt V, Poljak M et al. Acute fatty liver of pregnancy—an underlying condition for herpes simplex type 2 fulminant hepatitis necessitating liver transplantation. Z Gastroenterol 2005; 43(5):451-454.
59. Khan AA, Habeeb A, Parveen N et al. Peritoneal transplantation of human fetal hepatocytes for the treatment of acute fatty liver of pregnancy: a case report. Trop Gastroenterol 2004; 25(3):141-143.
60. Remiszewski P, Pawlak J, Skwarek A et al. Orthotopic liver transplantation for acute liver failure resulting from "acute fatty liver of pregnancy". Ann Transplant 2003; 8(3):8-11.

61. Westbrook RH, Yeoman AD, Joshi D et al. Outcomes of severe pregnancy-related liver disease: refining the role of transplantation. Am J Transplant 2010; 10(11):2520-2526.
62. Tang W, Huang Z, Wang Y et al. Effect of plasma exchange on hepatocyte oxidative stress, mitochondria function, and apoptosis in patients with acute fatty liver of pregnancy. Artif Organs 2012.
63. Treem WR, Rinaldo P, Hale DE et al. Acute fatty liver of pregnancy and long-chain 3-hydroxy-acyl-coenzyme A dehydrogenase deficiency. Hepatology 1994; 19(2):339-345.
64. Isaacs JD Jr, Sims HF, Powell CK et al. Maternal acute fatty liver of pregnancy associated with fetal trifunctional protein deficiency: molecular characterization of a novel maternal mutant allele. Pediatr Res 1996; 40(3):393-398.
65. Eaton S, Bartlett K, Pourfarzam M. Mammalian mitochondrial beta-oxidation. Biochem J 1996; 320(Pt 2):345-357.
66. Kamijo T, Wanders RJ, Saudubray JM et al. Mitochondrial trifunctional protein deficiency. Catalytic heterogeneity of the mutant enzyme in two patients. J Clin Invest 1994; 93(4):1740-1747.
67. Carpenter K, Pollitt RJ, Middleton B. Human liver long-chain 3-hydroxyacyl-coenzyme A dehydrogenase is a multifunctional membrane-bound beta-oxidation enzyme of mitochondria. Biochem Biophys Res Commun 1992; 183(2):443-448.
68. Uchida Y, Izai K, Orii T et al. Novel fatty acid beta-oxidation enzymes in rat liver mitochondria. II. Purification and properties of enoyl-coenzyme A (CoA) hydratase/3-hydroxyacyl-CoA dehydrogenase/3-ketoacyl-CoA thiolase trifunctional protein. J Biol Chem 1992; 267(2):1034-1041.
69. Jackson S, Kler RS, Bartlett K et al. Combined enzyme defect of mitochondrial fatty acid oxidation. J Clin Invest 1992; 90(4):1219-1225.
70. Izai K, Uchida Y, Orii T et al. Novel fatty acid beta-oxidation enzymes in rat liver mitochondria. I. Purification and properties of very-long-chain acyl-coenzyme A dehydrogenase. J Biol Chem 1992; 267(2):1027-1033.
71. Weinberger MJ, Rinaldo P, Strauss AW et al. Intact alpha-subunit is required for membrane-binding of human mitochondrial trifunctional beta-oxidation protein, but is not necessary for conferring 3-ketoacyl-CoA thiolase activity to the beta-subunit. Biochem Biophys Res Commun 1995; 209(1):47-52.
72. Ushikubo S, Aoyama T, Kamijo T et al. Molecular characterization of mitochondrial trifunctional protein deficiency: formation of the enzyme complex is important for stabilization of both alpha- and beta-subunits. Am J Hum Genet 1996; 58(5):979-988.
73. Aoyama T, Wakui K, Orii KE et al. Fluorescence in situ hybridization mapping of the alpha and beta subunits (HADHA and HADHB) of human mitochondrial fatty acid beta-oxidation multienzyme complex to 2p23 and their evolution. Cytogenet Cell Genet 1997; 79(3-4):221-224.
74. Orii KE, Orii KO, Souri M et al. Genes for the human mitochondrial trifunctional protein alpha- and beta-subunits are divergently transcribed from a common promoter region. J Biol Chem 1999; 274(12):8077-8084.
75. Kanazawa M, Ohtake A, Abe H et al. Molecular cloning and sequence analysis of the cDNA for human mitochondrial short-chain enoyl-CoA hydratase. Enzyme Protein 1993; 47(1):9-13.
76. Miyazawa S, Osumi T, Hashimoto T. The presence of a new 3-oxoacyl-CoA thiolase in rat liver peroxisomes. Eur J Biochem 1980; 103(3):589-596.
77. DiMauro S, DiMauro PM. Muscle carnitine palmityltransferase deficiency and myoglobinuria. Science 1973; 182(4115):929-931.
78. Vianey-Liaud C, Divry P, Gregersen N et al. The inborn errors of mitochondrial fatty acid oxidation. J Inherit Metab Dis 1987; 10 Suppl 1:159-200.
79. Tanaka K, Yokota I, Coates PM et al. Mutations in the medium chain acyl-CoA dehydrogenase (MCAD) gene. Hum Mutat 1992; 1(4):271-279.
80. Wood JC, Magera MJ, Rinaldo P et al. Diagnosis of very long chain acyl-dehydrogenase deficiency from an infant's newborn screening card. Pediatrics 2001; 108(1):E19.
81. Wanders RJ, Duran M, Ijlst L et al. Sudden infant death and long-chain 3-hydroxyacyl-CoA dehydrogenase. Lancet 1989; 2(8653):52-53.
82. Wanders RJ, L IJ, Poggi F et al. Human trifunctional protein deficiency: a new disorder of mitochondrial fatty acid beta-oxidation. Biochem Biophys Res Commun 1992; 188(3):1139-1145.

83. Aoyama T, Uchida Y, Kelley RI et al. A novel disease with deficiency of mitochondrial very-long-chain acyl-CoA dehydrogenase. Biochem Biophys Res Commun 1993; 191(3):1369-1372.
84. Bertrand C, Largilliere C, Zabot MT et al. Very long chain acyl-CoA dehydrogenase deficiency: identification of a new inborn error of mitochondrial fatty acid oxidation in fibroblasts. Biochim Biophys Acta 1993; 1180(3):327-329.
85. Boles RG, Buck EA, Blitzer MG et al. Retrospective biochemical screening of fatty acid oxidation disorders in postmortem livers of 418 cases of sudden death in the first year of life. J Pediatr 1998; 132(6):924-933.
86. Sinclair-Smith C, Dinsdale F, Emery J. Evidence of duration and type of illness in children found unexpectedly dead. Arch Dis Child 1976; 51(6):424-429.
87. Bennett MJ, Allison F, Pollitt RJ et al. Fatty acid oxidation defects as causes of unexpected death in infancy. Prog Clin Biol Res 1990; 321:349-364.
88. Bennett MJ, Powell S. Metabolic disease and sudden, unexpected death in infancy. Hum Pathol 1994; 25(8):742-746.
89. Pollitt RJ. Defects in mitochondrial fatty acid oxidation: clinical presentations and their role in sudden infant death. Padiatr Padol 1993; 28(1):13-17.
90. Ibdah JA, Tein I, Dionisi-Vici C et al. Mild trifunctional protein deficiency is associated with progressive neuropathy and myopathy and suggests a novel genotype-phenotype correlation. J Clin Invest 1998; 102(6):1193-1199.
91. L IJ, Ruiter JP, Hoovers JM et al. Common missense mutation G1528C in long-chain 3-hydroxyacyl-CoA dehydrogenase deficiency. Characterization and expression of the mutant protein, mutation analysis on genomic DNA and chromosomal localization of the mitochondrial trifunctional protein alpha subunit gene. J Clin Invest 1996; 98(4):1028-1033.
92. Hagenfeldt L, Venizelos N, von Dobeln U. Clinical and biochemical presentation of long-chain 3-hydroxyacyl-CoA dehydrogenase deficiency. J Inherit Metab Dis 1995; 18(2):245-248.
93. Brackett JC, Sims HF, Rinaldo P et al. Two alpha subunit donor splice site mutations cause human trifunctional protein deficiency. J Clin Invest 1995; 95(5):2076-2082.
94. L IJ, Wanders RJ, Ushikubo S et al. Molecular basis of long-chain 3-hydroxyacyl-CoA dehydrogenase deficiency: identification of the major disease-causing mutation in the alpha-subunit of the mitochondrial trifunctional protein. Biochim Biophys Acta 1994; 1215(3):347-350.
95. Wanders RJ, L IJ, van Gennip AH et al. Long-chain 3-hydroxyacyl-CoA dehydrogenase deficiency: identification of a new inborn error of mitochondrial fatty acid beta-oxidation. J Inherit Metab Dis 1990; 13(3):311-314.
96. Rinaldo P, Raymond K, al-Odaib A et al. Clinical and biochemical features of fatty acid oxidation disorders. Curr Opin Pediatr 1998; 10(6):615-621.
97. Pons R, Roig M, Riudor E et al. The clinical spectrum of long-chain 3-hydroxyacyl-CoA dehydrogenase deficiency. Pediatr Neurol 1996; 14(3):236-243.
98. Schaefer J, Jackson S, Dick DJ et al. Trifunctional enzyme deficiency: adult presentation of a usually fatal beta-oxidation defect. Ann Neurol 1996; 40(4):597-602.
99. Bertini E, Dionisi-Vici C, Garavaglia B et al. Peripheral sensory-motor polyneuropathy, pigmentary retinopathy, and fatal cardiomyopathy in long-chain 3-hydroxy-acyl-CoA dehydrogenase deficiency. Eur J Pediatr 1992; 151(2):121-126.
100. Przyrembel H, Jakobs C, L IJ et al. Long-chain 3-hydroxyacyl-CoA dehydrogenase deficiency. J Inherit Metab Dis 1991; 14(5):674-680.
101. Dionisi-Vici C, Garavaglia B, Burlina AB et al. Hypoparathyroidism in mitochondrial trifunctional protein deficiency. J Pediatr 1996; 129(1):159-162.
102. Ibdah JA, Dasouki MJ, Strauss AW. Long-chain 3-hydroxyacyl-CoA dehydrogenase deficiency: variable expressivity of maternal illness during pregnancy and unusual presentation with infantile cholestasis and hypocalcaemia. J Inherit Metab Dis 1999; 22(7):811-814.
103. Amirkhan RH, Timmons CF, Brown KO et al. Clinical, biochemical, and morphologic investigations of a case of long-chain 3-hydroxyacyl-CoA dehydrogenase deficiency. Arch Pathol Lab Med 1997; 121(7):730-734.
104. Tyni T, Palotie A, Viinikka L et al. Long-chain 3-hydroxyacyl-coenzyme A dehydrogenase deficiency with the G1528C mutation: clinical presentation of thirteen patients. J Pediatr 1997; 130(1):67-76.

105. Miyajima H, Orii KE, Shindo Y et al. Mitochondrial trifunctional protein deficiency associated with recurrent myoglobinuria in adolescence. Neurology 1997; 49(3):833-837.
106. Yang Z, Zhao Y, Bennett MJ et al. Fetal genotypes and pregnancy outcomes in 35 families with mitochondrial trifunctional protein mutations. Am J Obstet Gynecol 2002; 187(3): 715-720.
107. Tyni T, Ekholm E, Pihko H. Pregnancy complications are frequent in long-chain 3-hydroxyacyl-coenzyme A dehydrogenase deficiency. Am J Obstet Gynecol 1998; 178(3):603-608.
108. Ibdah JA, Perlegas P, Zhao Y et al. Mice heterozygous for a defect in mitochondrial trifunctional protein develop hepatic steatosis and insulin resistance. Gastroenterology 2005; 128(5):1381-1390.
109. Propst A, Propst T, Judmaier G et al. Prognosis in nonalcoholic steatohepatitis. Gastroenterology 1995; 108(5):1607.
110. Hultcrantz R, Glaumann H, Lindberg G et al. Liver investigation in 149 asymptomatic patients with moderately elevated activities of serum aminotransferases. Scand J Gastroenterol 1986; 21(1):109-113.
111. Barycki JJ, O'Brien LK, Bratt JM et al. Biochemical characterization and crystal structure determination of human heart short chain L-3-hydroxyacyl-CoA dehydrogenase provide insights into catalytic mechanism. Biochemistry (Mosc) 1999; 38(18):5786-5798.
112. Engel CK, Kiema TR, Hiltunen JK et al. The Crystal Structure of Enoyl-CoA Hydratase Complexed with Octanoyl-CoA Reveals the Structural Adaptations Required for Binding of a Long Chain Fatty Acid-CoA Molecule. J Mol Biol 1998; 275(5):859-847.
113. Spiekerkoetter U, Sun B, Khuchua Z et al. Molecular and phenotypic heterogeneity in mitochondrial trifunctional protein deficiency due to beta-subunit mutations. Hum Mutat 2003; 21(6):598-607.
114. Schwab KO, Ensenauer R, Matern D et al. Complete deficiency of mitochondrial trifunctional protein due to a novel mutation within the beta-subunit of the mitochondrial trifunctional protein gene leads to failure of long-chain fatty acid beta-oxidation with fatal outcome. Eur J Pediatr 2003; 162(2):90-95.
115. Orii KE, Aoyama T, Wakui K et al. Genomic and mutational analysis of the mitochondrial trifunctional protein beta-subunit (HADHB) gene in patients with trifunctional protein deficiency. Hum Mol Genet 1997; 6(8):1215-1224.
116. Ibdah JA, Zhao Y, Viola J et al. Molecular prenatal diagnosis in families with fetal mitochondrial trifunctional protein mutations. J Pediatr 2001; 138(3):396-399.
117. Shekhawat P, Bennett MJ, Sadovsky Y et al. Human placenta metabolizes fatty acids: implications for fetal fatty acid oxidation disorders and maternal liver diseases. Am J Physiol Endocrinol Metab 2003; 284(6):E1098-E1105.
118. Oey NA, den Boer ME, Ruiter JP et al. High activity of fatty acid oxidation enzymes in human placenta: implications for fetal-maternal disease. J Inherit Metab Dis 2003; 26(4):385-392.
119. Innes AM, Seargeant LE, Balachandra K et al. Hepatic carnitine palmitoyltransferase I deficiency presenting as maternal illness in pregnancy. Pediatr Res 2000; 47(1):43-45.
120. Ylitalo K, Vanttinen T, Halmesmaki E et al. Serious pregnancy complications in a patient with previously undiagnosed carnitine palmitoyltransferase 1 deficiency. Am J Obstet Gynecol 2005; 192(6):2060-2062.
121. Matern D, Hart P, Murtha AP et al. Acute fatty liver of pregnancy associated with short-chain acyl-coenzyme A dehydrogenase deficiency. J Pediatr 2001; 138(4):585-588.
122. Nelson J, Lewis B, Walters B. The HELLP syndrome associated wiht fetal medium-chain acyl-CoA dehydrogenase deficiency. J Inherit Metab Dis 2000; 23(5):518-519.
123. Santos L, Patterson A, Moreea SM et al. Acute liver failure in pregnancy associated with maternal MCAD deficiency. J Inherit Metab Dis 2007; 30(1):103.
124. Chakrapani A, Olpin S, Cleary M et al. Trifunctional protein deficiency: three families with significant maternal hepatic dysfunction in pregnancy not associated with E474Q mutation. J Inherit Metab Dis 2000; 23(8):826-834.
125. Browning MF, Levy HL, Wilkins-Haug LE et al. Fetal fatty acid oxidation defects and maternal liver disease in pregnancy. Obstet Gynecol 2006; 107(1):115-120.

SECTION IV
Liver Transplantation

Chapter 8

Pregnancy and Liver Transplantation

Ghassan M. Hammoud,* Abhishek Choudhary and Jamal A. Ibdah

Abstract

Women with decompensated liver disease are usually infertile and conception, if occurred, carries a high risk of maternal and fetal morbidity and mortality. Liver transplantation has improved the lives of patients with decompensated acute and chronic liver disease. With orthotopic liver transplantation, menstruation resumes and most women of childbearing age are able to conceive, give birth and lead a better quality of life. However, pregnancy in liver transplant recipients is challenging and considered high risk with increased incidence of prematurity, preterm labor, intrauterine growth restriction and cesarean delivery. To prevent rejection, liver transplant recipients are maintained on a life-long immunosuppressive regimen that must also be continued through pregnancy. With close monitoring and follow-up, most female liver transplant recipients of childbearing age will lead a successful pregnancy.

Introduction

Pregnancy is a complex physiologic condition and in women with transplant organs there are several controversies that encompass risks to the mother. These risks include allograft dysfunction and potential fetal and neonatal morbidity associated with immunosuppressive medication. Pregnancy in the setting of liver transplantation is considered high risk. However, pregnancy in the setting of cirrhosis is an uncommon condition possibly secondary to hepatocellular damage and altered sex hormones metabolism by the liver which lead to anovulation.[1,2]

During normal pregnancy there is an increase in maternal blood volume leading to an increase in portal venous pressure with some changes in liver functions.[3-6] Pregnant women with decompensated liver disease are at risk for bleeding associated with portal hypertension.[7] Liver failure occurring in pregnancy is rare and often caused by a liver disease specific to pregnancy such as acute fatty liver of pregnancy (AFLP), eclampsia-related liver disease, and the syndrome of Hemolysis, Elevated Liver enzymes, and Low Platelets (HELLP).[8,9] See Table 1.

Liver transplantation has been performed for over four decades now and is an acceptable therapy for many patients with acute liver failure and end-stage liver disease. Moreover, one-third of all liver transplant recipients are females and 75% of female recipients are of reproductive age.[10] In 2010, the expectation of 1-year post transplant survival is greater than 85% for most indications. Improved survival and quality of life following liver transplantation are associated with an increased frequency of pregnancies in liver-transplanted women. In this chapter we will review the impact on pregnancy during and after liver transplantation.

*Corresponding Author: Ghassan M. Hammoud—Department of Internal Medicine, Division of Gastroenterology and Hepatology, University of Missouri, Columbia, Missouri, USA. Email: hammoudg@health.missouri.edu

Maternal Liver Disease, edited by Jamal A. Ibdah. ©2012 Taylor & Francis Group, LLC

Table 1. Common causes of acute liver failure in pregnancy

First Trimester (Conception-12 Weeks)	Second Trimester (12-24 Weeks)	Third Trimester (24-40 Weeks)
Viral hepatitis (A, B, C, D, **E**)	Viral hepatitis (A, B, C, D, **E**)	Viral hepatitis (A, B, C, D, **E**)
Drug-induced hepatotoxicity	Drug-induced hepatotoxicity	Drug-induced hepatotoxicity
		Herpes simplex, cytomegalovirus, and Epstein Barr viruses
		Acute fatty liver of pregnancy
		The syndrome of Hemolysis, Elevated Liver enzymes, and Low Platelets (HELLP)
		Pre-eclempsia/eclempsia
		Budd-Chiari syndrome
		Acute hepatic rupture
		Autoimmune hepatitis
		Hereditary porphyria
		Wilson disease

Pregnancy before Liver Transplantation

Women with Decompensated Liver Disease

Pregnancy in the setting of chronic liver disease is a rare event and most women with decompensated chronic liver disease are infertile.[11,12] The incidence of cirrhosis in pregnancy is estimated to be approximately 1 in 5,950 pregnancies.[13] Women with alcoholic liver disease enter menopause at a significantly younger age than controls. Alcohol alters the hypothalamic-pituitary-ovarian (HPO) axis and ovarian function.[14,15]

Menstrual dysfunction in women with decompensated liver disease is common and menstrual abnormalities may be the first indication of liver disease. As many as 50% of women experience secondary amenorrhea, whereas others often experience oligomenorrhea, metrorrhagia and menorrhagia.[16-18] Hypothalamic-pituitary dysfunction seen in cirrhosis is associated with an inadequate response to GnRH agonists and clomiphene citrate as well as diminished gonadotrophin release relative to the reduced levels of circulating sex steroids.[15] Moreover, Estradiol level is increased in patients with portosystemic shunts.

The co-existence of pregnancy in patients with cirrhosis represents a complex clinical dilemma for the clinician. In portal hypertension (PHT), splanchnic blood flow and intrahepatic sinusoidal pressure increase leads to deviation of the blood to the azygos venous system with formation of collaterals.[19,20] This will result in elevated levels of estradiol and testosterone, potentially from portosystemic shunting of weak androgens which are converted to estrogens.[21]

Portal Hypertension

Pregnancy profoundly affects systemic hemodynamics particularly during the second and third trimesters. During pregnancy, a hypervolemic state develops leading to an increase in portal flow and elevation of portal venous pressure transmitted to the collateral veins with increased risk of esophageal variceal bleeding.[7,22,23] Portal hypertension occurs during the last stages of the second trimester of pregnancy and is associated with increased risk of esophageal variceal bleeding in the later stages of pregnancy. Esophageal variceal bleeding has been reported in 18 to 32% of pregnant women with cirrhosis and in up to 50% of those with known portal hypertension.[24,25] In those with pre-existing varices, 78% will have gastrointestinal bleeding during pregnancy leading to a mortality rate of 18 to 50%.[11] Pregnant patients with cirrhosis face unique risks. These include higher rates of spontaneous abortion, prematurity, pulmonary hypertension, splenic artery aneurysm rupture, postpartum hemorrhage, and a potential for life-threatening variceal hemorrhage and hepatic decompensation.[11,26,27] Indeed, the rates of cesarean section, preterm labor, peripartum infection and hypertension are higher than the general obstetrical population.[28,29] Shaheen and Myers reported in a large population-based study (conducted from 1993 to 2005) the outcomes of pregnancy in 339 patients with cirrhosis.[29] The authors found maternal and fetal mortality were much higher than the general population [(1.8% vs 0%, $P < 0.0001$); 5.2% vs 2.1%, $P < 0.0001$)], respectively. Moreover, gestational hypertension, placental abruption and uterovaginal hemorrhage were more common in patients with cirrhosis. Furthermore, the rate of hepatic decompensation occurred in 15% and patients with cirrhosis were more likely to deliver by cesarean delivery (42% vs 28%; adjusted odds ratio 1.41; 95% confidence interval 1.06-1.88). Similarly, the spontaneous abortion rate in cirrhotic patients is approximately 15 to 20%.[30]

Although cirrhosis is the most common cause of portal hypertension in the US, noncirrhotic portal hypertension (NCPH) represents an important cause in developing countries but with favorable outcome to pregnant women and their babies.[31] Esophageal variceal bleeding is the most common complication in pregnancies with NCPH occurring in 34% of cases, but the majority (88.2%) respond to endoscopic sclerotherapy.[32] Pregnancy outcome may be influenced by the underlying etiology of liver disease. Pregnant cirrhotic patients with underlying diagnosis of autoimmune hepatitis and alcoholic liver disease tend to have a favorable outcome in comparison to those with viral hepatitis.

Management of Esophageal Varices in Pregnancy

The management of a parturient with esophageal varices and thrombocytopenia require multidisciplinary care. Endoscopic surveillance and banding of esophageal varices is recommended during pregnancy. Upper endoscopy in general appears to be safe during pregnancy, with the main risk being fetal hypoxia from sedative drugs or positioning.[33] No cases of premature labor or fetal malformations have been reported in patients who have undergone endoscopy during pregnancy.[33]

On the basis of the endoscopic findings, primary prophylaxis with nonselective beta-blockers such as propranolol or nadolol (designated by the Food and Drug Administration as pregnancy category C) is recommended (Table 5).[34] The risks of nonselective beta-blockers include fetal bradycardia, hypotension, hypoglycemia and intrauterine growth restriction (IUGR).[35] Endoscopic banding ligation seems to be a safe procedure in pregnancy.[36] When bleeding is not arrested endoscopically, an emergency transjugular intrahepatic portosystemic shunt (TIPS) should be considered, but data in pregnant cirrhotic women are scarce.[37,38] Ascites rarely occurs in pregnant women with cirrhosis and the mainstay of treatment is sodium restriction and the use of diuretics.[39,40]

Hepatic Encephalopathy in Pregnancy

Hepatic encephalopathy in pregnancy may be related to a wide variety of acute and chronic liver diseases, including acute viral hepatitis, acute fatty liver of pregnancy, and preeclampsia-related liver injury.[41,42] In the setting of cirrhosis, pregnant women can develop portosystemic hepatic encephalopathy secondary to medications, sepsis, pancreatitis, renal failure, hypoxia, gastrointestinal bleeding and hypotension.[43,44] Treatment with lactulose and antibiotics such as rifaximin is effective (Table 2).[45,46]

Table 2. Management of portal hypertension in pregnancy

Findings	Prevalence	Management
Esophageal varices	18-32%	Primary prophylaxis: Large varices-nonselective β-blockers Secondary prophylaxis: Endoscopic band ligation and/sclerotherapy. TIPS*
Ascites	11%	Salt restriction, diuretics
Portosystemic encephalopathy	1.2%	Lactulose, rifaximin
Jaundice	Na	Therapy is based on etiology

*TIPS = Transjugular portosystemic shunt. The data on TIPS and pregnancy is scarce and controversial.

Postpartum Uterine Hemorrhage

Postpartum uterine hemorrhage occurs in 7 to 10% of pregnancies in patients with cirrhosis and represents a potential source of maternal morbidity and mortality. Vaginal delivery is usually safe and early assistance with forceps delivery or vacuum extraction should be considered to prevent further rise in portal venous pressure secondary to prolonged straining during labor. Termination of pregnancy is warranted in the presence of progressive hepatic decompensation.

Preconception Counseling

Pregnancy and Immunosuppression

Female liver transplant recipients can safely undergo pregnancy, although there are reports of high rates of premature birth and low birth weight infants particularly if pregnancy occurred in the first few months after transplantation. Pregnancy should be discouraged in the setting of decompensated chronic liver disease. The optimal timing of conception is a matter of debate but most transplant centers recommend that pregnancy be deferred at least 1 year following liver transplantation to minimize the risk of rejection and to avoid fetal exposure to high-dose immunosuppression. Liver transplant (LT) recipient women should be on a stable immunosuppressive regimen and have a stable graft function with no evidence of recent rejection at the time of pregnancy. The ideal condition would be to prepare the patient for conception once function of the liver and kidney is stabilized and patient is on a stable maintenance immunosuppressive agents. Once stable graft function has been achieved and maintenance regimens of immunosuppressive agents have been initiated, it is reasonable for women of childbearing age to consider pregnancy. Timing of pregnancy and immunosuppression in the liver transplant patient will be discussed in more detail below.

Counseling for Risks

Anticipated mothers should be counseled regarding risks of viral hepatitis transmission to their partners and their babies if pregnancy is planned. Women with viral hepatitis have an increased risk for complications during pregnancy.[47] Vertical hepatitis C virus (HCV) transmission rates range from 2-8%, with maternal viremia defined as detectable HCV RNA in blood is an almost uniform prerequisite for transmission.[48,49] In pregnancies among HCV-infected mothers who were HCV RNA negative, vertical transmission was rare.[49] Maternal co-infection with HIV significantly increases the risk of vertical HCV transmission to as much as 44%. In a recent cohort study, risk factors associated with increased rate of vertical HCV transmission were higher maternal HCV viral titer, prolonged membrane rupture during labor (6 hours or longer), and use of internal fetal monitoring during labor.[50] The risk of transmission through amniocentesis appears to be low for women who are chronically infected, although the number of exposed cases in the literature is small.

Contrary to hepatitis C virus, hepatitis B virus (HBV) has a high rate of vertical transmission causing fetal and neonatal hepatitis. Moreover, infant hepatitis B infection increases risk of chronic infection, cirrhosis or hepatocellular carcinoma in adult.[51] Because HBV is highly pathogenic and infectious, perinatal transmission of HBV infection represents the single largest cause of chronically infected individuals worldwide.[52] Approximately 10-20% of women who are seropositive for hepatitis B surface antigen (HBsAg) alone transmit the virus to their neonates in the absence of neonatal immunoprophylaxis. However, if the mother is seropositive for both HBsAg and hepatitis B e-antigen (HBeAg), the frequency of vertical transmission increases to approximately 90% without neonatal prophylaxis.[53] Therefore the American College of Obstetrics and Gynecology and the Centers for Disease Control (CDC) recommend universal screening for HBsAg of all pregnant women during the third trimester.[51] Prophylactic HBV immunoglobulin and hepatitis B vaccination is the current standard of care for HBV seropositive pregnant women to prevent HBV transmission to the infant.

Contraception

Female transplant recipients should also be counseled on the methods of contraception as each method has its own risks and benefits. Effective contraception is essential for the first 1-2 years after liver transplantation. There is a consensus that contraception should begin soon after transplantation, as menstruation can occur as early as 1 month posttransplant.[54] Contraception in liver transplant patients will be discussed in more detail below.

Breastfeeding

Counseling regarding the method of breastfeeding post liver transplantation is also warranted. The risk of transmission of viral hepatitis B infection through breast milk is very low and mothers should not discontinue breastfeeding if there is no evidence of broken breast-skin barrier.[55] Similarly, both HCV antibodies and HCV RNA have been detected in mothers' colostrums, however, there is little evidence that HCV is transmitted by breast milk; therefore, HCV-infected mothers need not avoid breastfeeding.

Comorbidities

Assessment of renal, cardiovascular, and pulmonary status should also be performed. Comorbidities such as diabetes and hypertension should be assessed and ideally be well controlled. Patients with genetic disorders notably alpha-one antitrypsin deficiency, hemochromatosis, and less commonly Allagille syndrome and Caroli syndrome should be offered genetic counseling. Although pregnancy outcomes after liver transplantation are frequently favorable, patients should be counseled regarding the possibility of maternal and fetal morbidity in these pregnancies. In addition, patients should be counseled regarding the possibility of adoption as a means to avoid the maternal and graft related complications a pregnancy may pose.

Timing of Pregnancy

Optimally, timing of pregnancy should be individualized for each patient. Although it is estimated that 25 to 50% of reproductive age women with end-stage liver disease suffer from infertility after liver transplantation, studies have demonstrated that normal reproductive function resume in 50 to 75% of premenopausal women after liver transplantation. In fact, pregnancies have been reported as early as 1 month after liver transplantation.[54,56] It is ideal to hold pregnancies till renal and liver functions are stabilized and the liver transplant patient is on a maintenance immunosuppression. Theoretically, delaying pregnancy 1 to 2 years after transplantation minimizes fetal exposure to high doses of immunosuppressants. In general, immunosuppressive agents are at their nadir approximately one year post liver transplantation with low risk of allograft rejection. Indeed, most recipients who experience rejection do so within the first 42 days after liver transplant.[57]

Medical clearance for pregnancy requires a team consultation with transplant hepatologist, maternal-fetal medicine specialist and other specialists. The current recommendation by most transplant centers is at least 1-year interval between transplantation and pregnancy.[58,59] By that time viral prophylaxis treatment has been completed in the first year, patients are on a maintenance immunosuppression medication, have a controlled blood pressure, and have no evidence of graft rejection. Female liver transplant recipients should also have a social support to be able to care for themselves and their infants. Therefore, it is important that the anticipated pregnant patient undergoes psychosocial evaluation to ensure optimum care for her and her future infant.

Contraception

According to data from the Organ Procurement and Transplantation Network (OPTN), 85,694 women underwent transplantation in the United States between January 2000 and September 2008 with 10,460 women receiving an organ transplant in 2007 alone. Of the nearly 86,000 transplants, 45% (38,741) were women between the ages of 18 and 49 years.[60] Additionally, kidney and liver are the most commonly transplanted organs. Within 1 year after liver transplantation, 97% of female graft recipients of childbearing age recover menstrual cycles and have the chance to conceive and give birth to a child. In a national cross-sectional study of all pediatric recipients who underwent liver transplantation between 1987 and 2007, 57 participants (86%) were compared to randomly chosen healthy controls (n = 141) at 10.7 ± 6.6 years posttransplant. In adult survivors, health-related quality of life (HRQoL) and sexual health are achievable post liver transplantation.[61] Since pregnancy in this population poses substantial maternal and fetal risks, discussion with the female liver transplant recipients about contraception is prudent. It is also imperative that liver transplant patients use safe and highly effective methods of contraception for prevention of unplanned pregnancy. Pregnancy should be avoided if possible during the first 12-24 months post liver transplantation.[62] Effective contraception should be initiated as early as three weeks posttransplantation as fertility resumes. The American Society of Transplantation (AST) recommends that obstetricians and gynecologists discuss posttransplant contraception with the transplant recipients of reproductive age before they undergo liver transplant procedure.

The preferred method of contraception during this period has been of a debate. Barrier methods have been recommended giving their low inherent risks of systemic complications although they may carry a high failure rates. Barrier methods are generally recommended during the first few months post transplantation since they don't pose any risk to the mother, if any, besides their benefit of reducing incidence of sexually transmitted disease.[63] Intrauterine devices (IUDs) use has been generally discouraged in posttransplant setting due to their potential infection complications associated with immunosuppression and their failure rates since they rely on macrophages for their mechanism of action.[63] However, immunosuppressive agents are thought to have minimal effect on macrophage activity. Data on the use of IUDs in nontransplant immunosuppressed patients such as HIV-positive patients provided no significant increase in complications.[64] Given the limited data available at this time, IUDs use post liver transplantation can't be recommended.

Hormonal combined oral contraceptives (COCs) are known to be effective and relatively well tolerated. A retrospective study by Jabiry-Zieniewicz et al to assess the tolerability and safety of low-dose hormonal contraceptives among 15 female liver transplant recipients, aged 24 to 35 years followed for 12 months, found no negative impact on liver function or glucose metabolism within 6 months to 7 years post transplantation.[65] Blood pressure and BMI remained stable and oral contraceptives provided 100% efficacy. There was no case of graft rejection or thromboembolic phenomena post transplantation.

Although combined oral contraceptives are known to be effective and relatively well tolerated, they are not devoid of some side effects such as breast tenderness, dyspepsia, nausea, vaginal staining, headache, weight gain and venous thromboembolism.[65] A recent systematic review by Paulen et al to evaluate evidence on the safety and effectiveness of contraceptive use among women having undergone solid organ transplantation concluded that there was limited evidence on combined oral contraceptives (COCs) and transdermal contraceptive patch use among kidney and liver transplant recipients with no overall changes in biochemical measures.[60] It is also worth noting that hormonal contraceptives do possess other noncontraceptive benefits such as regularity of menses, diminished blood loss, and decrease/lack of dysmenorrhea and symptoms of premenstrual syndrome. On the contrary, their use is associated with increased risk of venous thromboembolism.[66] Therefore, the use of combined oral contraceptives in patients with history of deep venous thrombosis and pulmonary embolism is contraindicated. Moreover, their risks outweigh their benefits in patients with hypertension. Indeed, 75% of liver transplant recipients develop hypertension post liver transplantation. A report from the American Society of Transplantation's consensus conference on reproduction and transplantation concluded that there is no information to suggest that combined oral contraceptives are associated with adverse consequences among hypertensive transplant patients, when hypertension is well controlled.[67]

The efficacy of combined oral contraceptives may be affected with the concomitant use of immunosuppressive medication considering that both are metabolized by the enzyme cytochrome P450 3A4 (CYP3A4). Glucocorticoids are inducers of CYP34A and therefore the efficacy of combined oral contraceptives may be decreased. Conversely, inhibition of CYP3A4 by COCs could either increase or decrease the plasma level of immunosuppressive medications such as tacrolimus, cyclosporine and sirolimus.[60,68] Vaginal administration seems to be a safe route to achieve steady and precise doses of contraceptive hormones, resulting in stable serum concentrations and low exposure.[69] The estrogen containing contraceptive vaginal ring may diminish the chance of drug interactions and therefore be safer for patients.[69] Therefore, all liver transplant recipient women receiving hormonal contraceptives should undergo close monitoring of liver function and immunosuppressant levels in the blood to detect early deterioration of graft function and determine whether such deterioration is related to hormone therapy or caused by an impaired graft function due to a primary pretransplant disease.

Pregnancy Outcome in Female Liver Transplant Recipients

The first successful human liver transplant was performed in 1967 by Dr. Thomas Starzl at the University of Colorado. Since then, advances in surgical techniques, immunosuppression, and patient selection have made the procedure a standard life-saving treatment for patient with decompensated acute and chronic liver disease. Eleven years later, the first successful pregnancy in a liver graft recipient was reported by Walcott and colleagues in 1978. Given the improving success of liver transplantation over the past two decades and decreasing levels of immunosuppression, most organ transplant recipients lead happy and healthy lives with an average 1-year survival rate of >85% at better transplant programs.

Based on the data from the USA National Transplantation Pregnancy Registry (NTPR), >70% of posttransplant pregnancies have a successful livebirth. Although pregnancy after liver transplantation does not seem to have obviously deleterious effects on allograft function or survival, complications can occur both in mothers and infants. Most outcome data on pregnancy during and after liver transplantation are obtained from The National Transplantation Pregnancy Registry which was established in 1991 at Thomas Jefferson University in Philadelphia, Pennsylvania, to study the outcomes of pregnancies in transplant recipients in North America, including female transplant recipients and those fathered by male transplant recipients. In 1997, the United Kingdom Transplant Registry was established to collect information on the outcomes of pregnancies of transplant patients in the United Kingdom. Both of these registries shed the light on our understanding of the outcome of pregnancy during and after liver transplantation (Table 3).

Fetal Complications

In 1978, Walcott and coworkers documented the first pregnancy in a liver transplant recipient. The recipient delivered a healthy male, weighing 2400 gm at term.[70] In 1991, the National Transplantation Pregnancy Registry was established with the aim of collecting and analyzing data about pregnancies and deliveries among graft recipients. As of 2010, there were more than 2000 pregnancy outcomes in female transplant recipients reported to the National Transplantation Pregnancy Registry.[71] For the NTPR's largest group, female kidney recipients, 71-76% of the pregnancies produced a live birth. For the other organs combined, the livebirth likelihood ranges from 50-86%. The incidence of birth defects in the liveborn appears similar to the general population, except for pregnancies with mycophenolic acid (MPA) exposure that have a 23% incidence of birth defects.[71] In 131 liver transplant recipients in the NTPR, 219 pregnancies and 223 outcomes were reported. Children born to female liver recipients have a greater risk of prematurity and low birth weight than the general population; however, no congenital malformation patterns have been observed. Data among 205 pregnancies in LT recipients described in 2006 NTPR, 19% and 2% ended in spontaneous abortions and stillbirths, respectively.[59] It is presumed that low birth weight may be linked to placental vascular damage caused by hypertension or the immunosuppressive regimen. Moreover, pregnancies early after liver transplantation carries an increased risk of prematurity. In a small series of 71 pregnancies after liver transplantation from King's College Hospital, London, Christopher and coworkers compared outcomes before and after one year of transplantation in the recipients.[56] The authors found that there were no statistically significant differences in abortion and live birth rates between the early and late liver transplant recipient groups. However, there was a trend toward increased rates of nearly all adverse outcomes in the early group (within one year of transplant), notably prematurity (58% versus 29%) and rejection (33% versus 14%).

In a meta-analysis and review of the literature by Dei Malatesta and colleagues, they summarized the outcome of 285 pregnancies in liver transplant recipients.[72] The authors found a significantly shorter interval between transplant and conception in the unsuccessful pregnancies, comprised of spontaneous and therapeutic abortions compared to the successful pregnancies (24 ± 20 versus 48 ± 42, $P < 0.001$). The incidence of perinatal and neonatal death was approximately 4%. Moreover, there was a 3% incidence of congenital malformations. In the UK Transplant Pregnancy Registry, Sibanda and colleagues reported in 2007 the outcome data on maternal and fetal factors and pregnancy outcomes in solid organ transplant recipients between 1994 and 2001.[73] There were live births in 83%, 69%, and 79% of pregnancies in cardiothoracic organ, liver, and kidney recipients, respectively. In 16 liver transplant recipients, premature birth (<37 weeks) was seen in 50%, low birth weight (<2500 gm) in 57%, spontaneous abortion in 13% and therapeutic abortion in 13%. Of the 69% live births from liver patients, 62% of the preterm infants were delivered via cesarean section.[73] It is difficult to determine whether immunosuppressive therapies lead to adverse pregnancy outcomes or whether these outcomes stem from the underlying maternal disease itself. Certainly, the use of immunosuppressive agents to prevent graft rejection has some effects on fetal maturity.

The fetus is invariably exposed to these medications during development. Maternal glucocorticoid therapy has been associated with an increased rate of preterm labor, premature rupture of the membranes and intrauterine growth restrictions. Calcineurin inhibitors such as cyclosporine and tacrolimus have been associated with congenital malformations and premature birth. A recent study by Jabiry-Zieniewicz and coworkers evaluated the outcomes in 39 pregnant liver

Table 3. Summary of important fetal and maternal outcomes in liver transplant recipients from selected publications

Author, Year, Number of Pregnancies, [Reference]	Live Birth Rate (%)	Preterm (%)	Graft Dysfunction (%)	Cesarean Section Rate (%)	Spontaneous Abortions (%)	Low Birthweight <2500 g (%)	Maternal/ Neonatal Deaths (%)
Nagy et al,[76] 2003 n = 38	63	29	17	46	Na	17	17/0
Jain et al,[78] 2003 n = 49	100	4	25	47	0	9	10/6
Armenti et al,[59] 2006 n = 205	73	35	7	35	19	34	0/0
Christopher et al,[56] 2006 n = 71	71	Na	17	40	19	20	4/0
Dei Malatesta et al,[72] 2006 n = 285	78	31	10	43	Na	23	0/4
Sibanda et al,[73] 2007 n = 16	69	50	Na	62	13	57	Na
Coffin et al,[79] 2010 n = 206	70	27	5	38	5	Na	0/6
Jabiry-Zieniewics et al,[74] 2011 n = 39	100	31	8	80	0	20	0/0

recipients' deliveries from 2001 to 2011 and reported the risk of prematurity, intrauterine growth restriction, and fetal distress at 30%, 20% and 15.4%, respectively.[74] See Table 3.

Sonographic confirmation of gestational age should be obtained early in the first trimester to provide accurate interpretation of fetal growth scans in the second and third trimesters. Moreover, given the increase incidence of low birth weight, preterm labor and intrauterine growth restriction, serial ultrasounds to assess fetal growth is advisable.

Maternal Complications

Pregnancies after liver transplantation are generally associated with good outcomes and seem to not affect graft and patient survival. However, high rates of obstetric complications can occur. Pregnancy in these patients is considered high risk and seems to be associated with increased health risks, including ectopic pregnancy, hypertension and pre-eclempsia, infection, and need for delivery by cesarean section.[75,76] Indeed, pregnant female liver transplant recipients should be monitored carefully. Moreover, maternal use of immunosuppressive medications may increase the risk of gestational complications. For example, cyclosporine (CSA) and glucocorticoids (GCs) are associated with hypertension, diabetes mellitus, lower mean birth weight and IUGR.

In a large single-center experience that evaluated the outcome of 71 pregnancies after liver transplantation, pregnancy induced hypertension was seen in 20%, preeclampsia in 13%, acute cellular rejection in 17%, and renal impairment in 11%.[56] Pregnancies occurring within 1-year posttransplantation had an increased incidence of prematurity (58% vs 29%), low birth weight, and acute cellular rejection (33% vs 14%) compared with those occurring later than 1 year. Although no statistically significant differences in outcome were identified, rates of acute rejection and graft loss are similar to nonpregnant liver recipients.The NTPR registry published in 2006 reported a 7% rate of acute rejection episodes during pregnancy and 8% graft loss within 2 years after delivery.[59] Moreover, the NTPR analysis of liver recipients noted that those who reported rejection during or after pregnancy had poorer newborn outcomes and graft survival.[77] If desired, pregnancy should be deferred for 1 year after liver transplantation. This recommendation is supported by NTPR data.

It is difficult to determine the impact of the primary cause of liver failure on pregnancy outcomes. In a study that evaluated 31 pregnancies in 16 women with autoimmune hepatitis as their primary diagnosis, it was concluded that pregnancy outcomes were comparable with those of recipients with other diagnoses and that pregnancy had no negative impact on maternal survival in this group.[59] Review of NTPR data for 13 maternal deaths in liver transplant recipients showed the causes of death are commonly related to recurrent liver disease post liver transplantation, liver failure, renal failure, sepsis, multiorgan dysfunction, cardiac arrest and portal vein thrombosis. The median interval for death was 43 (6-156) months.[75] A recent study by Jabiry-Zieniewicz and coworkers evaluated the maternal outcome, in 39 pregnant liver recipients from 2001 to 2011.[74] 26 LT recipients were on tacrolimus as the main immunosuppressive agent while 9 LT recipients received cyclosporine as the main immunosuppressive agent. The most common obstetric complications were premature labor (30.8%), hypertension (25.6%), symptomatic urinary tract infections (18%), preeclampsia (7.7%), premature rupture of the membranes (7.7%), cholestasis (7.7%) and severe anemia (5.51%). The mean time from liver transplantation to delivery was 67.6 months. The rate of acute graft rejection was 7.7% of the liver transplant recipients and 79.5% of LT recipients had cesarean section as the mode of delivery (Table 3). The most common indication for delivery by cesarean section was fetal distress, prolonged labor, previous cesarean section and breech presentation. It is noteworthy that none of the LT recipients experienced graft loss within 2 years after delivery, although only 10 liver transplant recipients remained in 2-year follow-up.[74] The authors did not observe increase incidence of congenital malformations. These data confirm that pregnancies after liver transplantation are possible and may complete successfully.

Mode of Delivery in Women after Liver Transplantation

Orthotopic liver transplantation (OLT) results in restored fertility of patients of reproductive age; so female liver transplant recipients have the chance to conceive, give birth, lead normal lives and continue professional careers. The patient status after liver transplantation does not determine the way of delivery and it is not by itself considered to be an indication for cesarean delivery. Vaginal delivery appears to be safe and should always be the aim. Cesarean section is indicated in the setting of obstetrical complications such as preeclampsia, breech presentation, fetal distress, imminent intrauterine infection, premature labor, failed vaginal delivery and previous cesarean delivery.

There is an increased rate of delivery by cesarean section in conception that occurred in the first year post liver transplantation. Of the 69% live births from liver recipient patients, 62% of the preterm infants delivered via cesarean.[73] It has also been reported that the rate of cesarean section deliveries among liver transplant recipients is higher than that of the general population.[78,79] In fact, fetal distress is the major cause for cesarean section in these women. Jabiry-Zieniewicz and coworkers retrospectively evaluated the mode of delivery in 17 liver recipients with 21 deliveries from 2001 to 2006.[80] Most patients were primigravidas on tacrolimus-based immunosuppressive regimens. The mean time from transplantation to pregnancy was 4.3 ± 3.6 years. The authors found that high rate of cesarean sections (71%) in liver transplant recipients are associated with a great incidence of obstetric complications of pregnancy. All neonates were delivered in a good state with a mean birth weight of 2725 gm. A case-controlled study reported by Coffin et al from the Healthcare Cost and Utilization Project Nationwide Inpatient Sample (NIS) database from 1993 to 2005 identified 146 delivery admissions among liver transplant recipients.[79] The authors found that cesarean deliveries were more common among transplant recipients (38% versus 24%; $P = 0.0001$). Data from the NTPR in 2006 reported a 35% cesarean section rate on 121 females liver transplant recipients.[81] Nagy et al reported a 45.8% cesarean delivery rate among liver recipients.[76] These data signify the high rate of cesarean delivery in liver transplant recipients. It is not known whether a certain immunosuppressive regimen is associated with an increase rate of cesarean delivery.

Immunosuppressants and Liver Transplant in Pregnancy

Immunosuppressive medications are the mainstay of the medical management of liver transplant recipients. Although there is no consensus on the optimal maintenance regimen for transplant recipients, immunosuppressive agents in liver transplantation typically include calcineurin inhibitors (CNI) such as cyclosporine (CSA) or tacrolimus (TAC), antimetabolites such as mycophenolate mofetil (MMF) with or without azathioprine (AZA) and the mammalian target of rapamycin (mTOR) inhibitors (either sirolimus or everolimus), and prednisone. Additional medications for liver transplant recipients may include prophylaxis for pneumocystis carinii infection with trimethoprim/sulfamethoxazole or pentamidine, and for herpes virus infection with acyclovir.

Most immunosuppressive agents (e.g., azathioprine, cyclosporine, and MMF) have been found to be teratogenic in animals. However, the frequency of birth defects in infants born to women receiving immunosuppressive medication is not statistically different from that in the general population.[82] Caution must be exercised when interpreting animal and observational data given that teratogenicity may be species-dependent.

All immunosuppressive medication cross the placenta and enter fetal circulation raising questions about the long-term outcomes in children exposed in utero. Theoretically, this could result in severe immunosuppressive state in the newborn. The altered immune function in infants exposed to immunosuppressive agents in pregnancy generally resolves and normalizes within the first year of life. This transient immune dysfunction can affect the baby's immunoreactive response to immunization. This potentially could put the infant at risk for illnesses in the neonatal period especially if live vaccination is administered.

Long-term data on the developmental outcomes of infants born to solid organ recipients are limited although some data suggested a 16-24% incidence of developmental delay in children between the ages of 5 and 12 years. Patients treated with either cyclosporine or tacrolimus should have serial blood tests in pregnancy to follow the drug levels and to assess hepatic and renal functions while avoiding unnecessary toxicity. Family planning and multidisciplinary team involvement are optimum.

Classification of Immunosuppression Risk in Pregnancy

The Food and Drug Administration (FDA) has developed a classification system that codifies the safety categories of immunosuppressive medications in pregnancy (Table 4). Medications are designated as either category A, B, C, D, or X (Table 5). This classification is difficult to apply in the setting of immunosuppression during pregnancy, particularly since many of the data are derived from experimental and preclinical animal studies.

Table 4. The food and drug administration pregnancy categories

Pregnancy Category	Definition
A	Controlled studies show no risk.
B	Animal studies show no risk, and there are no human controlled studies, or animal studies may show an adverse effect that was not reproduced in human controlled studies.
C	Risk cannot be ruled out; human studies are lacking and either animal studies show an adverse effect, or there are no studies available.
D	Positive evidence of risk in human studies; use may be considered if the potential benefits outweigh the risk.
X	Contraindicated in pregnancy: Risk is evident in animal and human studies and outweighs any benefit.

Table 5. Summary of the food and drug administration pregnancy categories of immunosuppression medication

Drug	Pregnancy Category
Tacrolimus	C
Cyclosporine	C
Mycophenolate mofetil	D
Sirolimus	C
Corticosteroids	B
Antithymocyte globulin	C
Azathioprine	D
Propranolol	C
Nadolol	C

Data collected in transplanted women treated with immunosuppressive drugs has determined no consistent pattern of congenital malformations in the offspring and the documented prevalence of major malformations has been found to be 4-5%, similar to the 3% documented in healthy women.[83]

Glucocorticoids (FDA Category B)

Animal studies as well as human studies have shown increased risk of cleft palate in the offspring of women treated with glucocorticoids. Park-Wyllie et al in a meta-analysis showed a marginally increased risk of major malformations after first-trimester exposure to corticosteroids.[84] The authors found that prednisone does not represent a major teratogenic risk in humans at therapeutic doses, however, it does increase by an order of 3.4-fold the risk of oral cleft, which is consistent with the existing animal studies.

Maternal glucocorticoids therapy has also been associated with adrenal insufficiency in the fetus, increase rate of premature rupture of the membranes and intrauterine growth restriction possibly due to the effect of glucocorticoid on the pituitary-adrenal axis. Moreover, glucocorticoid treatment is known to result in transient hypertension and hyperglycemia. Moreover, chronic glucocorticoids therapy has been shown to lead to immunosuppression, impaired wound healing with increased incidence of hematomas and skin breakdown.[85]

Azathioprine (FDA Categories D) and Mycophenolate (FDA Categories D)

Azathioprine and mycophenolate are both antimetabolites frequently used in conjunction with corticosteroids as part of maintenance immunosuppression regimen post transplantation. Azathioprine is an inhibitor of purine metabolism and after absorption is rapidly converted in the liver into a number of metabolites. Azathioprine crosses the placenta but remains in its inactive form. Although it is not converted to the known teratogen 6-mercaptopurine in the fetus, animal data has demonstrated structural and hematological abnormalities in rodents. Similarly, in humans, in utero exposure of azathioprine has been associated with reports of congenital malformations, including hypospadias and preaxial polydactyly.

Maternal immunosuppression with azathioprine during pregnancy has also been associated with isolated cases of immunoglobulin deficiency and fetal immunosuppression. Azathioprine can depress fetal haemopoiesis resulting in neonatal thrombocytopenia and leucopenia with the potential for serious outcomes. In the mid-1990s, mycophenolate mofetil (MMF) was introduced, which belongs to the anti-metabolite class of immunosuppressives, and this agent largely replaced azathioprine.

MMF, a category D agent, is a prodrug that undergoes hepatic ester hydrolysis to form the active metabolite mycophenolic acid, a reversible inhibitor of inosine monophosphate dehydrogenase, which blocks de novo purine synthesis on which lymphocytes are dependent. MMF has been associated with a number of structural fetal malformations such as hypoplastic nails, shortened fifth fingers, abnormal ears, cleft lip/palate and agenesis of the corpus callosum. The current recommendations do not support the use of MMF in pregnant transplant recipients.

Cyclosporine and Tacrolimus (FDA Categories C)

Calcineurin inhibitors such as cyclosporine and tacrolimus are considered the mainstay of immunosuppressive therapy post transplantation and are not used together. Tacrolimus is a macrolide antibiotic approved in the US in 1995, which is a more potent calcineurin inhibitor than cyclosporine. Data from the NTPR revealed 4 to 5% prevalence of fetal structural anomalies in pnegnancies exposed to either cyclosporine or tacrolimus. There was no specific pattern of malformation detected. Animal data suggested increased risk of abnormal renal development, however no human data supported this risk.

Sirolimus (FDA Category C)

The mammalian target of rapamycin (mTOR) inhibitor, sirolimus, is a macrolide antibiotic and is used in maintenance immunosuppression regimens. Sirolimus, a category C agent, approved in the US in 1999, has no effect on calcineurin activity but inhibits cytokine driven T-cell proliferation. Sirolimus possesses no renal side effects and potentially protects against the development of malignancy as a renal sparing medication. Data on 39 pregnancies who have reported pregnancies to the National Transplantation Pregnancy Registry indicated no structural malformation in children of mothers exposed to sirolimus. The current recommendations don't support its use in pregnancy given its potential teratogenic potential.

Breastfeeding by Mothers Receiving Immunosuppressive Medication

Breastfeeding for transplant recipient mothers remains an evolving area. Breastfeeding is generally not recommended during immunotherapy, based on theoretical concerns of affecting the development of immune function in the baby. However, there is a lack of reported adverse effects in the literature. Tacrolimus is one of the main immunosuppressant drug used orally post liver transplantation to prevent graft rejection and is considered FDA pregnancy category C. Tacrolimus exhibits variable but usually low absorption from the gastrointestinal tract, and is extensively bound to proteins (99%); the whole-blood-to-plasma ratio is approximately 20. These factors suggest that there is low excretion of the drug into human breast milk and a low risk for absorption by the suckling infant. Case reports suggest that infant exposure to tacrolimus in milk is very low and that maternal tacrolimus therapy may be compatible with breastfeeding. It is worth noting that infants exposed to immunosuppressants may have lower levels of B cells, T cells, and immunoglobulins.[86] Furthermore, infants exposed to immunosuppressants may have inadequate response to vaccines and possibly develop overt illnesses.[87]

Similarly, infant's blood cyclosporine concentration was usually very low in female patients of childbearing age who are receiving cyclosporine. Breastfed infants of mothers treated with CSA received less than 300 microgram per day of CSA and absorbed undetectable amounts (Nyberg, transplantation 1998). Data collected from the NTPR indicated no adverse outcomes in infants who were breastfed during maternal cyclosporine use.

Mycophenolate mofetil and azathioprine are considered pregnancy category D and should not be used in pregnancy. Azathioprine is excreted in breast milk and metabolized to 6-mercaptopurine. Traditionally, women receiving azathioprine have been discouraged from breastfeeding because of theoretical potential risks of neonatal bone marrow suppression, susceptibility to infection, and pancreatitis. However, there is insufficient evidence in the literature to suggest that women taking azathioprine should refrain from breastfeeding.

Prednisone is excreted in breast milk and its concentration peak in breast milk occurs within the first 2 hours after ingestion. Therefore, it is recommended that mothers abstain from breastfeeding within the first 2 hours after ingestion. There are no adverse effects have been demonstrated in infants breastfed in the setting of maternal glucocorticoid treatment. The American Academy of Pediatrics concluded that prednisone and other glucocorticoids can safely be used by lactating mothers.

Conclusion

Pregnancies are occurring in liver transplant recipients with increasing frequency. Successful pregnancy outcomes have been described after liver transplantation, although the timing of pregnancy is critical. Considering the rapid return of fertility following transplantation, accurate family planning advice and counseling is essential. Although the majority of pregnancies in liver transplant recipients have favorable outcomes, both mothers and their infants have an increased risk of complications. To date there is no evidence of specific structural malformations among children born to female liver transplant recipients, but there appears to be an ever-present risk of prematurity and/or low birth weight. Future studies with large sample size are needed to assess both the short- and long-term consequences related to the use of immunosuppressive medications in pregnancy and their effects on germ cell lines during development.

References

1. Borhanmanesh F, Haghighi P. Pregnancy in patients with cirrhosis of the liver. Obstet Gynecol 1970; 36(2):315-324.
2. Benjaminov FS, Heathcote J. Liver disease in pregnancy. Am J Gastroenterol 2004; 99(12): 2479-2488.
3. Munnell EW, Taylor HC Jr. Liver blood flow in pregnancy; hepatic vein catheterization. J Clin Invest 1947; 26(5):952-956.
4. Rovinsky JJ, Jaffin H. Cardiovascular hemodynamics in pregnancy. 3. Cardiac rate, stroke volume, total peripheral resistance, and central blood volume in multiple pregnancy. Synthesis of results. Am J Obstet Gynecol 1966; 95(6):787-794.
5. Bacq Y, Zarka O, Brechot JF et al. Liver function tests in normal pregnancy: a prospective study of 103 pregnant women and 103 matched controls. Hepatology 1996; 23(5):1030-1034.
6. Lund CJ, Donovan JC. Blood volume during pregnancy. Significance of plasma and red cell volumes. Am J Obstet Gynecol 1967; 98(3):394-403.
7. d'Alteroche L, Perarnau JM, Perrotin F et al. Portal hypertension and pregnancy. Gastroenterol Clin Biol 2008; 32(5 Pt 1):541-546.
8. Sandhu BS, Sanyal AJ. Pregnancy and liver disease. Gastroenterol Clin North Am 2003; 32(1):407-436, ix.
9. Chen H, Yuan L, Tan J, Liu Y et al. Severe liver disease in pregnancy. Int J Gynaecol Obstet 2008; 101(3):277-280.
10. Rinella ME. Pregnancy after liver transplantation. Ann Hepatol 2006; 5(3):212-215.
11. Tan J, Surti B, Saab S. Pregnancy and cirrhosis. Liver Transpl 2008; 14(8):1081-1091.
12. Viner RM, Forton JT, Cole TJ et al. Growth of long-term survivors of liver transplantation. Arch Dis Child 1999; 80(3):235-240.
13. Aggarwal N, Sawnhey H, Suril V et al. Pregnancy and cirrhosis of the liver. Aust N Z J Obstet Gynaecol 1999; 39(4):503-506.
14. Hugues JN, Perret G, Adessi G et al. Effects of chronic alcoholism on the pituitary-gonadal function of women during menopausal transition and in the post menopausal period. Biomedicine 1978; 29(8):279-283.
15. Van Thiel DH, Lester R. The effect of chronic alcohol abuse on sexual function. Clin Endocrinol Metab 1979; 8(3):499-510.
16. Douglas NC, Shah M, Sauer MV. Fertility and reproductive disorders in female solid organ transplant recipients. Semin Perinatol 2007; 31(6):332-338.
17. Stellon AJ, Williams R. Increased incidence of menstrual abnormalities and hysterectomy preceding primary biliary cirrhosis. Br Med J (Clin Res Ed) 1986; 293(6542):297-298.
18. Cundy TF, Butler J, Pope RM et al. Amenorrhoea in women with non-alcoholic chronic liver disease. Gut 1991; 32(2):202-206.
19. Garcia-Tsao G. Portal hypertension. Curr Opin Gastroenterol 2005; 21(3):313-322.
20. Minano C, Garcia-Tsao G. Clinical pharmacology of portal hypertension. Gastroenterol Clin North Am 2010; 39(3):681-695.
21. Gordon GG, Olivo J, Rafil F et al. Conversion of androgens to estrogens in cirrhosis of the liver. J Clin Endocrinol Metab 1975; 40(6):1018-1026.
22. Britton RC. Pregnancy and esophageal varices. Am J Surg 1982; 143(4):421-425.
23. Lopez-Mendez E, Avila-Escobedo L. Pregnancy and portal hypertension a pathology view of physiologic changes. Ann Hepatol 2006; 5(3):219-223.
24. Britton RC. Pregnancy and esophageal varices. Am J Surg 1982; 143(4):421-425.
25. Pajor A, Lehoczky D. Pregnancy and extrahepatic portal hypertension. Review and report on the management. Gynecol Obstet Invest 1990; 30(4):193-197.

26. Sigel CS, Harper TC, Thorne LB. Postpartum sudden death from pulmonary hypertension in the setting of portal hypertension. Obstet Gynecol 2007; 110(2 Pt 2):501-503.
27. Safioleas MC, Moulakakis KG. A rare cause of intra-abdominal haemorrhage: spontaneous rupture of the splenic vein. Acta Chir Belg 2006; 106(2):237-239.
28. Murthy SK, Heathcote EJ, Nguyen GC. Impact of cirrhosis and liver transplant on maternal health during labor and delivery. Clin Gastroenterol Hepatol 2009; 7(12):1367-1372, 72 e1.
29. Shaheen AA, Myers RP. The outcomes of pregnancy in patients with cirrhosis: a population-based study. Liver Int 2010; 30(2):275-283.
30. Pajor A, Lehoczky D. Pregnancy in liver cirrhosis. Assessment of maternal and fetal risks in eleven patients and review of the management. Gynecol Obstet Invest 1994; 38(1):45-50.
31. Sumana G, Dadhwal V, Deka D et al. Non-cirrhotic portal hypertension and pregnancy outcome. J Obstet Gynaecol Res 2008; 34(5):801-804.
32. Aggarwal N, Sawhney H, Vasishta K et al. Non-cirrhotic portal hypertension in pregnancy. Int J Gynaecol Obstet 2001; 72(1):1-7.
33. O'Mahony S. Endoscopy in pregnancy. Best Pract Res Clin Gastroenterol 2007; 21(5):893-899.
34. Tripathi D, Hayes PC. Review article: a drug therapy for the prevention of variceal haemorrhage. Aliment Pharmacol Ther 2001; 15(3):291-310.
35. Cissoko H, Jonville-Bera AP, Swortfiguer D et al. Neonatal outcome after exposure to beta adrenergic blockers late in pregnancy. Arch Pediatr 2005; 12(5):543-547.
36. Ghidirim G, Mishin I, Dolghii A et al. Prophylactic endoscopic band ligation of esophageal varices during pregnancy. J Gastrointestin Liver Dis 2008; 17(2):236-237.
37. Lodato F, Cappelli A, Montagnani M et al. Transjugular intrahepatic portosystemic shunt: a case report of rescue management of unrestrainable variceal bleeding in a pregnant woman. Dig Liver Dis 2008; 40(5):387-390.
38. Savage C, Patel J, Lepe MR et al. Transjugular intrahepatic portosystemic shunt creation for recurrent gastrointestinal bleeding during pregnancy. J Vasc Interv Radiol 2007; 18(7):902-904.
39. Jacobs WM, White L. Ascites in Pregnancy. Obstet Gynecol 1965; 25:872-874.
40. Reiss A, Lowenstein L, Solt I et al. Recurrent ascites due to oral contraceptives and pregnancy. Obstet Gynecol 2006; 107(2 Pt 2):499-501.
41. Easton JD. Severe preeclampsia/eclampsia: hypertensive encephalopathy of pregnancy? Cerebrovasc Dis 1998; 8(1):53-58.
42. de Nagy J, Strittmatter CA, Lee RV. Acute fatty liver of pregnancy presenting as hepatic encephalopathy at 28 weeks' gestation. J Obstet Gynaecol 2008; 28(1):100-102.
43. Davies MH, Wilkinson SP, Hanid MA et al. Acute liver disease with encephalopathy and renal failure in late pregnancy and the early puerperium—a study of fourteen patients. Br J Obstet Gynaecol 1980; 87(11):1005-1014.
44. Al Inizi S, Robertson IG. Acute hepatic encephalopathy with acute pancreatitis in pregnancy. J Obstet Gynaecol 2006; 26(3):257-258.
45. Aggarwal R. Hepatic encephalopathy in pregnancy. Indian J Gastroenterol 2003; 22 Suppl 2:S78-S80.
46. Russell MA, Craigo SD. Cirrhosis and portal hypertension in pregnancy. Semin Perinatol 1998; 22(2):156-165.
47. Reddick KL, Jhaveri R, Gandhi M et al. Pregnancy outcomes associated with viral hepatitis. J Viral Hepat 2011; 18(7):e394-e398.
48. Tajiri H, Miyoshi Y, Funada S et al. Prospective study of mother-to-infant transmission of hepatitis C virus. Pediatr Infect Dis J 2001; 20(1):10-14.
49. Airoldi J, Berghella V. Hepatitis C and pregnancy. Obstet Gynecol Surv 2006; 61(10):666-672.
50. Mast EE, Hwang LY, Seto DS et al. Risk factors for perinatal transmission of hepatitis C virus (HCV) and the natural history of HCV infection acquired in infancy. J Infect Dis 2005; 192(11):1880-1889.
51. Degli Esposti S, Shah D. Hepatitis B in pregnancy: challenges and treatment. Gastroenterol Clin North Am 2011; 40(2):355-372, viii.
52. Fontana RJ. Hepatitis B: a "GLOBAL" health challenge. Gastroenterology 2009; 136(2):389-392.
53. Chang MH. Hepatitis B virus infection. Semin Fetal Neonatal Med 2007; 12(3):160-167.

54. Laifer SA, Darby MJ, Scantlebury VP et al. Pregnancy and liver transplantation. Obstet Gynecol 1990; 76(6):1083-1088.
55. Pol S, Corouge M, Fontaine H. Hepatitis B virus infection and pregnancy. Clin Res Hepatol Gastroenterol 2011; 35(10):618-622.
56. Christopher V, Al-Chalabi T, Richardson PD et al. Pregnancy outcome after liver transplantation: a single-center experience of 71 pregnancies in 45 recipients. Liver Transpl 2006; 12(7):1138-1143.
57. Wiesner RH, Demetris AJ, Belle SH et al. Acute hepatic allograft rejection: incidence, risk factors, and impact on outcome. Hepatology 1998; 28(3):638-645.
58. Mastrobattista JM, Katz AR. Pregnancy after organ transplant. Obstet Gynecol Clin North Am 2004; 31(2):415-428, vii.
59. Armenti VT, Daller JA, Constantinescu S et al. Report from the National Transplantation Pregnancy Registry: outcomes of pregnancy after transplantation. Clin Transpl 2006; 57-70.
60. Paulen ME, Folger SG, Curtis KM et al. Contraceptive use among solid organ transplant patients: a systematic review. Contraception 2010; 82(1):102-112.
61. Kosola S, Lampela H, Lauronen J et al. General health, health-related quality of life and sexual health after pediatric liver transplantation: A nationwide study. Am J Transplant 2011.
62. Riely CA. Contraception and pregnancy after liver transplantation. Liver Transpl 2001; 7(11 Suppl 1):S74-S76.
63. Estes CM, Westhoff C. Contraception for the transplant patient. Semin Perinatol 2007; 31(6):372-377.
64. Stringer EM, Kaseba C, Levy J et al. A randomized trial of the intrauterine contraceptive device vs hormonal contraception in women who are infected with the human immunodeficiency virus. Am J Obstet Gynecol 2007; 197(2):144 e1-e8.
65. Jabiry-Zieniewicz Z, Bobrowska K, Kaminski P et al. Low-dose hormonal contraception after liver transplantation. Transplant Proc 2007; 39(5):1530-1532.
66. Samuelsson E, Hagg S. Incidence of venous thromboembolism in young Swedish women and possibly preventable cases among combined oral contraceptive users. Acta Obstet Gynecol Scand 2004; 83(7):674-681.
67. McKay DB, Josephson MA, Armenti VT et al. Reproduction and transplantation: report on the AST Consensus Conference on Reproductive Issues and Transplantation. Am J Transplant 2005; 5(7):1592-1599.
68. Gaston RS. Maintenance immunosuppression in the renal transplant recipient: an overview. Am J Kidney Dis 2001; 38(6 Suppl 6):S25-S35.
69. Paternoster DM, Riboni F, Bertolino M et al. The contraceptive vaginal ring in women with renal and liver transplantation: analysis of preliminary results. Transplant Proc 2010; 42(4):1162-1165.
70. Walcott WO, Derick DE, Jolley JJ et al. Successful pregnancy in a liver transplant patient. Am J Obstet Gynecol 1978; 132(3):340-341.
71. Coscia LA, Constantinescu S, Moritz MJ et al. Report from the National Transplantation Pregnancy Registry (NTPR): outcomes of pregnancy after transplantation. Clin Transpl 2010; 65-85.
72. Dei Malatesta MF, Rossi M, Rocca B et al. Pregnancy after liver transplantation: report of 8 new cases and review of the literature. Transpl Immunol 2006; 15(4):297-302.
73. Sibanda N, Briggs JD, Davison JM et al. Pregnancy after organ transplantation: a report from the UK Transplant pregnancy registry. Transplantation 2007; 83(10):1301-1307.
74. Jabiry-Zieniewicz Z, Szpotanska-Sikorska M, Pietrzak B et al. Pregnancy outcomes among female recipients after liver transplantation: further experience. Transplant Proc 2011; 43(8):3043-3047.
75. Armenti VT, Herrine SK, Radomski JS et al. Pregnancy after liver transplantation. Liver Transpl 2000; 6(6):671-685.
76. Nagy S, Bush MC, Berkowitz R et al. Pregnancy outcome in liver transplant recipients. Obstet Gynecol 2003; 102(1):121-128.
77. Armenti VT, Radomski JS, Moritz MJ et al. Report from the National Transplantation Pregnancy Registry (NTPR): outcomes of pregnancy after transplantation. Clin Transpl 2000; 123-134.
78. Jain AB, Reyes J, Marcos A et al. Pregnancy after liver transplantation with tacrolimus immunosuppression: a single center's experience update at 13 years. Transplantation 2003; 76(5):827-832.

79. Coffin CS, Shaheen AA, Burak KW et al. Pregnancy outcomes among liver transplant recipients in the United States: a nationwide case-control analysis. Liver Transpl 2010; 16(1):56-63.
80. Jabiry-Zieniewicz Z, Bobrowska K, Pietrzak B et al. Mode of delivery in women after liver transplantation. Transplant Proc 2007; 39(9):2796-2799.
81. Armenti VT, Constantinescu S, Moritz MJ et al. Pregnancy after transplantation. Transplant Rev (Orlando) 2008; 22(4):223-240.
82. Armenti VT, Radomski JS, Moritz MJ et al. Report from the National Transplantation Pregnancy Registry (NTPR): outcomes of pregnancy after transplantation. Clin Transpl 2002:121-130.
83. McKay DB, Josephson MA. Pregnancy in recipients of solid organs—effects on mother and child. N Engl J Med 2006; 354(12):1281-1293.
84. Park-Wyllie L, Mazzotta P, Pastuszak A et al. Birth defects after maternal exposure to corticosteroids: prospective cohort study and meta-analysis of epidemiological studies. Teratology 2000; 62(6):385-392.
85. Anstead GM. Steroids, retinoids, and wound healing. Adv Wound Care 1998; 11(6):277-285.
86. Hollander GA, Bierer BE, Burakoff SJ. Molecular and biological actions of cyclosporin A and FK506 on T-cell development and function. Transfus Sci 1994; 15(3):207-220.
87. Cimaz R, Meregalli E, Biggioggero M et al. Response to tetanus vaccination in infants exposed in utero to immunosuppressants for maternal autoimmune disorders. Lupus 2007; 16(2):129-132.

Index

I

J

L

M

N

O

P

R

S

T

U

V

W

For Product Safety Concerns and Information please contact our EU representative GPSR@taylorandfrancis.com Taylor & Francis Verlag GmbH, Kaufingerstraße 24, 80331 München, Germany

Printed and bound by CPI Group (UK) Ltd, Croydon, CR0 4YY
14/09/2026
02216223-0015